Evaluation of Miex® Process Impacts on Different Source Waters

Awwa
Research
Foundation

About the Awwa Research Foundation

The Awwa Research Foundation (AwwaRF) is a member-supported, international, nonprofit organization that sponsors research to enable water utilities, public health agencies, and other professionals to provide safe and affordable drinking water to consumers.

The Foundation's mission is to advance the science of water to improve the quality of life. To achieve this mission, the Foundation sponsors studies on all aspects of drinking water, including supply and resources, treatment, monitoring and analysis, distribution, management, and health effects. Funding for research is provided primarily by subscription payments from approximately 1,000 utilities, consulting firms, and manufacturers in North America and abroad. Additional funding comes from collaborative partnerships with other national and international organizations, allowing for resources to be leveraged, expertise to be shared, and broad-based knowledge to be developed and disseminated. Government funding serves as a third source of research dollars.

From its headquarters in Denver, Colorado, the Foundation's staff directs and supports the efforts of more than 800 volunteers who serve on the board of trustees and various committees. These volunteers represent many facets of the water industry, and contribute their expertise to select and monitor research studies that benefit the entire drinking water community.

The results of research are disseminated through a number of channels, including reports, the Web site, conferences, and periodicals.

For subscribers, the Foundation serves as a cooperative program in which water suppliers unite to pool their resources. By applying Foundation research findings, these water suppliers can save substantial costs and stay on the leading edge of drinking water science and technology. Since its inception, AwwaRF has supplied the water community with more than $300 million in applied research.

More information about the Foundation and how to become a subscriber is available on the Web at **www.awwarf.org**.

Evaluation of Miex® Process Impacts on Different Source Waters

Prepared by:
George C. Budd, **Bruce W. Long**, and **Jessica C. Edwards-Brandt**
Black & Veatch Corporation
8400 Ward Parkway, Kansas City, Missouri 64114
and
Philip C. Singer and **Maria Meisch**
Department of Environmental Sciences and Engineering
University of North Carolina, Chapel Hill, North Carolina

Jointly sponsored by:
Awwa Research Foundation
6666 West Quincy Avenue, Denver, CO 80235-3098

Southern Nevada Water Authority

Tampa Bay Water

Charleston Commissioners of Public Works

and

Northern Kentucky Water District

Published by:

Awwa Research Foundation
Advancing the Science of Water®

DISCLAIMER

This study was funded by the Awwa Research Foundation (AwwaRF), and co-funded by the Southern Nevada Water Authority, Tampa Bay Water, the Charleston Commissioners of Public Works, and the Northern Kentucky Water District. AwwaRF and the co-funding utilities assume no responsibility for the content of the research study reported in this publication or for the opinions or statements of fact expressed in the report. The mention of trade names for commercial products does not represent or imply the approval or endorsement of AwwaRF or the co-funding utilities. This report is presented solely for informational purposes.

CONTENTS

TABLES

FIGURES

FOREWORD

The Awwa Research Foundation is a nonprofit corporation that is dedicated to the implementation of research efforts to help utilities respond to regulatory requirements and traditional high-priority concerns of the industry. The research agenda is developed through a process of consultation with subscribers and drinking water professionals. Under the umbrella of a Strategic Research Plan, the Research advisory Council prioritizes the suggested projects based upon current and future needs, applicability, and past work; the recommendations are forwarded to the Board of Trustees for final selection. The foundation also sponsors research projects through the unsolicited proposal process; the Collaborative Research, Research Applications, and Tailored Collaboration programs; and various joint research efforts with organizations such as the U.S. Environmental Protection Agency, the U.S. Bureau of Reclamation, and the Association of California Water Agencies.

This publication is a result of one of these sponsored studies, and it is hoped that its findings will be applied in communities throughout the world. The following report serves not only as a means of communicating the results of the water industry's centralized research program but also as a tool to enlist the further support of the nonmember utilities and individuals.

Projects are managed closely from their inception to the final report by the foundation's staff and large cadre of volunteers who willingly contribute their time and expertise. The foundation serves a planning and management function and awards contracts to other institutions such as water utilities, universities, and engineering firms. The funding for this research effort comes primarily from the Subscription Program, through which water utilities subscribe to the research program and make an annual payment proportionate to the volume of water they deliver, and consultants and manufacturers subscribe based on their annual billings. The program offers a cost-effective and fair method for funding research in the public interest.

A broad spectrum of water supply issues is addressed by the foundation's research agenda: resources, treatment and operations, distribution and storage, water quality and analysis, toxicology, economics, and management. The ultimate purpose of the coordinated effort is to assist water suppliers to provide the highest possible quality of water economically and reliably. The true benefits are realized when the results are implemented at the utility level. The foundation's trustees are pleased to offer this publication as a contribution toward that end.

Water utilities are facing increasingly stringent water treatment regulations and public health goals for known contaminants. In order to meet tougher regulations, it is imperative that water utilities optimize unit processes. The use of a magnetic ion exchange process is a innovative approach to meeting these goals. The Foundation undertook this research project to provide guidance to utilities and consultants in evaluating the effectiveness, reliability, and costs associated with this innovative process for a variety of source water quality conditions.

Walter J. Bishop
Chair, Board of Trustees
Awwa Research Foundation

James F. Manwaring, P.E.
Executive Director
Awwa Research Foundation

ACKNOWLEDGMENTS

The authors of this report are indebted to the following water utilities and individuals for their cooperation and participation in this project:

- Maytum Water Treatment Plant, Tampa Bay Water
  - Christine Owen, John McKeon, George Witty
- Hanahan Water Treatment Plant, Charleston Commissioners of Public Works
  - John Cook, Dixie Fanning, Mark Cline, Peter Horlbeck, Ron Copeland
- Fort Thomas Water Treatment Plant, Northern Kentucky Water District
  - Bari Joslyn, Jim Johnson, Mary Carol Wagner, Bill Wulfeck
- Alfred Merritt Smith Water Treatment Facility, Southern Nevada Water Authority
  - Dave Rexing, Eric Wert, Glenn Gifford, Shane Snyder, SNWA laboratory

The advice to the project team and advocacy for the project provided by the Awwa Research Foundation Project Managers and thoughtful guidance of the Project Advisory Committee are gratefully acknowledged. My Linh Nguyen and Kenan Ozekin have provided much appreciated support and guidance as project officers. The Project Advisory Committee members are Karen Harr - Chesapeake Department of Public Utilities, Tanju Karanfil - Clemson University, Yu-Jung Chang - HDR Engineering, Chandra Mysore - US Filter/Vivendi.

The authors wish to acknowledge the immense generosity of the Water Corporation of Western Australia for their assistance in compiling information on the Wanneroo Groundwater Treatment Plant. The authors also wish to acknowledge Orica Corporation for their continual support and donation of resin for this study, including shipment costs and readiness to contribute to drinking water treatment research; and Westech Engineering, Inc., for their support in fabrication and operational assistance with the pilot plant.

EXECUTIVE SUMMARY

The MIEX® process can be operated as a series of mixed contactors to treat raw water without pretreatment, or it can be located at an intermediate point in the treatment train. The MIEX® resin is magnetized to facilitate recovery in a settling compartment downstream from the contactors. Most of the recovered resin is returned to the contactors, but a portion is regenerated to provide a steady state condition for maintaining ion exchange capacity in the system. A small fraction of fresh resin is added to compensate for losses. This project evaluates the performance of the MIEX® process under a wide range of conditions. Interaction with other plant elements that include disinfection and coagulation-based processes is another critical area that was assessed.

Core objectives for bench and pilot scale testing of the MIEX® process were to determine: (1) its effectiveness in removing natural organic matter (NOM) as measured by UV254 and dissolved organic carbon (DOC); (2) its capability to reduce the formation of trihalomethanes (THMs) and haloacetic acids (HAAs); (3), its effect on molecular weight size fractions of DOC and on hydrophilic/hydrophobic organic carbon distribution; (4) its effect on coagulant demand, the removal of bromide, and the formation of brominated organics, as well as on chlorine demand and ozone demand; and (5) resin capacity following multiple cycles of regeneration. Testing was also conducted at selected sites to assess removal of sulfide, perchlorate, and arsenic, all of which are anions that can potentially be removed by the MIEX® process.

Results from full-scale operation of the Wanneroo Groundwater Treatment Plant (WGTP) were combined with results from bench and pilot scale studies to develop additional guidance for planning and design. Bench and pilot-scale studies were conducted at four locations:

- Maytum Water Treatment Plant in New Port Richey, FL, a member of the Tampa Bay Water System.
- Hanahan Water Treatment Plant in Charleston, SC, operated by the Commissioners of Public Works of Charleston.
- Fort Thomas Water Treatment Plant in Fort Thomas, KY, operated by the Northern Kentucky Water District.
- Alfred Merritt Smith Water Treatment Facility in Las Vegas, NV, operated by the Southern Nevada Water Authority.

The Maytum Water Treatment Plant (MWTP) treats groundwater with aeration for sulfide removal and chlorination for disinfection, oxidation of sulfide, and reduction of color. MIEX® resin was tested at this location for its effectiveness in removing total organic carbon (TOC) for control of disinfection by-products to meet future regulations. The anion exchange properties of MIEX® resin also provide a potential for sulfide removal, which is of interest at this location.

The Hanahan Water Treatment Plant (HWTP) treats surface waters with high levels of TOC. Bromide is also a significant factor that affects disinfection by-product formation, and the potential for removal of bromide prior to disinfection is therefore important at this location. In addition, the effect of the MIEX® process on the downstream coagulation process is an important consideration, as NOM can have a significant effect on coagulant demand.

The Fort Thomas Water Treatment Plant (FTWTP) provides treatment of Ohio River water by conventional coagulation, flocculation, sedimentation, and filtration. The water quality of the Ohio River is characterized by moderate alkalinity and TOC concentrations, and periodically high turbidities. Testing was performed during a period of high rainfall; raw water turbidity during this time was as high as 150 NTU and averaged 78 NTU. The goals for evaluation of the MIEX® process at this location were primarily directed towards removing DBP precursors and enhancing the coagulation process.

The Alfred Merritt Smith Water Treatment Facility (AMSWTF) in Southern Nevada treats water from Lake Mead and serves the Las Vegas area. The source water is characterized by low turbidity and organics content, and is therefore treated with relatively low doses of ferric chloride as the primary coagulant in conjunction with a polymer for conditioning prior to direct filtration. The plant is being upgraded to incorporate a pre-ozonation step. Opportunities for enhanced coagulation through increased ferric chloride dosages are limited by constraints of the direct filtration process. In addition, the high alkalinity of the source water makes pH adjustment to enhance the removal of organics by coagulation difficult. Therefore, NOM removal capabilities are limited in the absence of the MIEX® process or other processes such as activated carbon adsorption or membrane treatment. Also, the capability of MIEX® resin for removing other anions such as bromide, arsenic, and perchlorate is a consideration at this location.

Each of these locations has different conditions under which the MIEX® process might be applied, thereby providing a broad basis for evaluation. Test results from all four sites have been compiled and evaluated for process applicability under different conditions. Specific findings from these sites in combination with observations at the full-scale WGTP are:

- Successful full-scale performance of the MIEX® process has been demonstrated since November, 2001 through operation of the 30 mgd (112.5 ML/day) WGTP. The plant treats a groundwater source that contains TOC typically greater than 10 mg/L. DOC removals of approximately 50 percent have been achieved. Average time of resin turnover within the plant is in excess of 1 year based on the observed resin loss rate and the total resin inventory. Therefore, there is extensive opportunity for exposure of the resin to potential fouling conditions. However, fouling has not proved to be a problem. The process operates in concert with a downstream alum coagulation step to achieve overall removals of NOM that can approach 75 percent.
- Testing at the four sites has indicated a wide range of susceptibility of NOM to removal. DOC removals were 60 percent or greater at the Maytum WTP (MWTP) and the Hanahan WTP (HWTP), while DOC removal at the Fort Thomas WTP (FTWTP) was slightly lower (50 to 60 percent removal), and removals of less than 35 percent were observed at the Alfred Merritt Smith WTP (AMSWTF). THM and HAA formation are reduced as a consequence of these DOC reductions, although the extent reduction proved to be site-specific. Therefore, applicability of the MIEX® process to meet specific goals may vary with location, and testing is recommended to confirm capability to meet target objectives.
- Testing conducted with the MIEX® process demonstrated the ability to provide significant improvements in downstream coagulation at the HWTP by reducing NOM and thus, the coagulant dose needed. The resulting reduction in coagulant with a corresponding reduction in residuals generated would reduce the costs for both chemicals and residuals disposal at this location. Coagulant reductions at the FTWP

were not observed for a dual coagulant (ferric sulfate combined with a commercial acidified alum polymer blend) approach that is presently applied at the plant, but significant reductions were observed for a single coagulant approach using ferric sulfate.

- The MIEX® process in combination with coagulation at the HWTP produced significantly greater reductions in DOC and UV254 than achieved by either process alone, which further lowered the formation of THMs and HAAs. Combining the MIEX®process and coagulation did not have a significant effect on DOC removal at FTWTP. However, there were differences in the extent of UV254 removal and significant reduction in the formations of THM's and HAA's were observed as compared with those achieved by coagulation alone. Therefore, some changes in the characteristics of the NOM were occurring although not reflected in a gross measure of DOC. On balance, the effect of MIEX® resin in combination with coagulation, although potentially beneficial, was not as significant at this location as at the HWTP and reflect a location-specific nature of this process sequence.
- Reductions in DOC through the MIEX® process will affect downstream oxidant demands. Reductions in chlorine demand were observed at all test locations. Evaluation of the effects of MIEX® pretreatment on ozonation indicated a reduced rate of ozone degradation at the AMSWTF. The slight increase in the initial ozone demand observed at this location was attributed to resin carryover from the MIEX® process. This effect is less likely to be of concern where MIEX® particulates can be removed by a clarification process before ozonation.
- Removal of selected inorganic anions (bromide, sulfide, arsenic, and perchlorate) tended to be somewhat variable, and dependent on both the nature of the anions and the water quality of the feed water. High source water TDS concentrations are indicative of background levels of anions that will compete more strongly with target anions for available ion exchange sites. Sulfate is of particular concern, because its divalent character results in a higher affinity for exchange sites than monovalent anions.
- Bromide removal was assessed at three of the testing locations. The most effective removal was achieved at the HWTP during a period that coincided with preparation of regenerant brine with a low bromide rock salt. In contrast, erratic removals were observed with brine prepared from a solar salt derived from a sea water source. Solar salt that is obtained from sea water can have significant levels of bromide that could interfere with restoration of the resin's sorptive capacity for bromide. Bromide removal was less effective at the AMSWTF even when a low bromide salt was used. The raw water supply to this plant contains high levels of TDS and particularly high levels of sulfate; therefore, there is more potential for anion interference with bromide uptake. It should be noted that the use of alternative salt sources for brine make up was not a specific objective of this study and that these results are not definitive. This aspect of the MIEX® process may merit further evaluation to follow up on these observations.
- Other inorganic anions that were evaluated in this study included sulfide, which was evaluated at the MWTP, and arsenic and perchlorate, which were evaluated at the AMSWTF. A high degree of sulfide removal was observed at the MWTP, whereas the removals of perchlorate and arsenic under high TDS conditions at the AMSWTF

were at lower levels. It is probable that the removal rates of arsenic and perchlorate will be site-specific, and that they could be removed more effectively from water supplies with lower TDS and sulfate concentrations. This type of application would, however, need to consider the potentially hazardous classification for disposal of the waste brine as consequence of the concentrating effects of the process.

- The MIEX® process causes some shifting of salt composition in the treated water. Sulfate is removed to some extent, while chloride is increased as a consequence of its exchange with anions that are being taken up by the resin. This exchange has some offsetting effects in terms of impact on the total salt content, and the TDS content of the treated water is not changed significantly.
- Key factors that affect process performance are resin concentration, contact time, and regeneration percentage. It should be noted that rate of resin addition to the process is not proportional to the resin concentration. In fact, observations at the full-scale WGTP suggest that resin addition to make up for losses may be decreased as the resin concentration maintained within the process increases, possibly as a result of more effective contacting and agglomeration of resin particles in the sedimentation process. For that reason, resin doses less than 10 mL/L were not examined in pilot testing in this study.
- Opinions of probable costs were prepared for a 10 mgd (37.5 ML/day) facility based on experience with construction and operation of the 30 mgd (112.5 ML/day) WGTP. An average annual flow of 6.7 mgd (25.1 ML/day) was used as the basis for developing operation and maintenance costs for this facility. Cost opinions were adjusted to reflect US construction price levels. The capital cost of this type of facility is estimated to be $ 3.9 million, based on a MIEX® concentration of 20 mL/L, a contact time of 15 minutes, and a 5 percent regeneration ratio. These conditions were selected as generally comparable to conditions that were identified at the four test sites in this study. Actual costs will be site-specific depending on bidding climate, local factors, and specifics of the process configuration. Operating and maintenance costs will vary depending on the rate of resin loss. Early operating experience at the WGTP indicated a resin loss of 3.8 L/ML, but losses have since been reduced to 3.0 L/ML. Probable annual operating and maintenance costs at this rate of resin loss are approximately $340,000. Efforts to reduce resin loss are being made both at the WGTP and elsewhere. An estimate based on a resin loss of 2 L/ML was also developed to indicate the impact if further reductions in resin loss could be achieved. Annual operating and maintenance costs at this rate of loss were estimated to be $250,000. Cost for disposal of waste brine would represent an additional cost and would vary from a minimal cost if disposal to a sanitary sewer were acceptable to higher costs for other options. While the process has a strong focus on minimizing the volume of waste, the disposal aspect of implementation can vary and could become an important component of the overall cost depending on the options that are available.
- Cost estimates were based on the assumption of minimal loss of resin capacity over time. There was little indication of progressive loss of resin capacity over the life of the MIEX® process at the WGTP in spite of the fact that the process is challenged by potential fouling resulting from upstream oxidation of significant levels of iron. Evaluations of resin activity over the 5 week periods of pilot plant operation at each

of the sites in this study also did not exhibit differences over time. While these observations were based on relatively short periods of resin life, they did tend to support the observed stability of capacity that has been observed at the WGTP. These observations may not be sufficient for drawing definitive conclusions for all locations; more information will become available as more full scale applications are implemented and it is still recognized that capability to maintain capacity may prove to be site specific. However, the resin shows promise for maintaining capacity based on information that is available to date.

Study results indicate that the MIEX® process can have a range of potential applications to meet site-specific needs. The costs of the process would generally be greater than those associated with treatment changes such as optimizing coagulation for greater removal of NOM in a conventional treatment sequence or conversion to chloramines as an alternative disinfection approach to limit the formation of THMs and HAAs. Many utilities may elect to consider these types of changes as first priorities before considering the MIEX® process. However, these approaches may not always be adequate. In addition, coagulation processes may not be available in all locations or may have constraints that limit capability for further optimization such as would be the case for many direct filtration facilities. There is also a degree of reluctance to rely on chloramines at some locations and this approach is not always available. Where applicable, the MIEX® process may prove to be less costly for NOM removal than approaches such as GAC or nanofiltration and could offer utilities a reasonable alternative to these technologies for situations where added levels of NOM removal are desired. Achievable levels of NOM removal may be greater with these other types of technologies and situations could occur where the MIEX® process may not provide the removals that are desired or where these technologies may prove to be more cost effective. However, the levels of NOM removal that are available with the MIEX® process may be appropriate at less cost to meet the needs of many utilities. Other alternatives that may become considerations in a larger disinfection and disinfection byproduct planning context are alternative oxidant/disinfection approaches, such as ozone, chlorine dioxide, and membrane filtration. In some cases, the MIEX® process might be used in combination with some of these processes. Therefore, a range of possibilities exist. Costs and process applicability of the respective process alternatives will vary depending on site specific factors. Costs that relate to disposal of regenerant in particular could exhibit significant variation for the MIEX® process from one site to the next. Benefits of the MIEX® process in relation to other plant processes will also be site specific. These benefits can include reductions in coagulant dose, solids to be disposed, and oxidant/disinfectant doses, and possible reductions in capital costs of downstream processes. These benefits will be considerations in computing the net cost impact of adding the MIEX® process.

The MIEX® process can achieve significant levels of removal of NOM and reductions in DBP formation, but the extent of these effects varies with water quality and it is important to confirm applicability with pilot testing. Bench scale testing can augment pilot testing, but caution should be exercised in using resin that has not been conditioned in a flow through sequence as the full-scale process relies on resin that has reached a steady state condition. Testing with virgin resin will not reflect removals that can be expected as a consequence. In many cases, the MIEX® process will be integrated within a treatment sequence and will provide a component of the overall strategy for NOM removal and DBP control. It is important that the process be examined for its capability to act in conjunction with other processes to meet the overall goals for NOM removal and DBP control under these conditions.

CHAPTER 1
INTRODUCTION

DESCRIPTION OF SITES

Water utilities are facing the challenge of having to provide higher levels of disinfection while reducing the concentrations of disinfection by-products. A key factor in meeting this challenge is the ability to remove naturally occurring organic matter (NOM). One emerging alternative for achieving removal of NOM is a magnetized ion exchange approach known as the MIEX® process, the subject of this report. This process has been demonstrated to be effective in removing dissolved organic carbon (DOC) from a range of waters, and has worked well in conjunction with coagulation by both reducing the dose of coagulant needed and by helping to remove NOM as characterized by DOC and UV254 absorbing materials.

The MIEX® process, developed specifically for the removal of NOM, utilizes a magnetized anion exchange resin. The structural and chemical characteristics of this micro-size, macroporous, strong base anion exchange resin make it highly resistant to organic fouling and allow it to be repeatedly regenerated. The resin contains quartenary amine functional groups that serve as the charge sites. Its internal specific surface area is comparable to that of conventional macroporous resins, but its small particle diameter of 150 to 180 μm gives it a large external surface area which benefits exchange kinetics and improves its resistance to fouling. In addition to these features, the resin incorporates a magnetic component in its polymeric structure, which makes individual resin beads behave like small magnets capable of forming large, heavy agglomerates that facilitate its recovery and reuse through several regeneration cycles. An additional property of the resin that is significant to water treatment is the high degree of physical stability that allows recycle pumping with minimal damage to the resin.

The magnetized anion exchange resin is applied continuously to a mixed tank where the ion exchange process occurs (see Figure 1.1). A resin dose of 10 to 20 mL/L is typically used to remove organics from water during a 10 to 30 minute detention time in a continuously stirred reactor; more recently, higher resin concentrations have been used successfully at full scale at the WGTP in Western Australia. The resin is recovered in a gravity settling basin with a relatively high overflow rate, and the treated water overflows to the next treatment step. Separation of the resin by gravity is very efficient because of the "magnetically" enhanced agglomeration of individual resin beads that produces agglomerates capable of settling at relatively high overflow rates. The settled resin is pumped back to the contactor as a concentrated slurry. A fraction of the recycled resin is continuously removed for regeneration and replaced with regenerated resin.

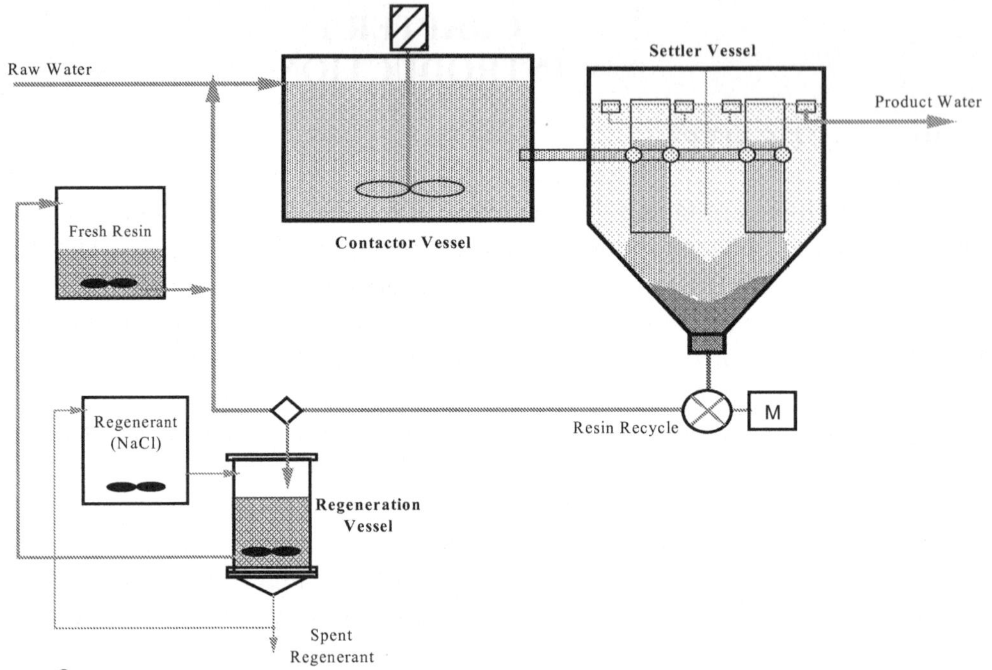

Figure 1.1 MIEX® Process Diagram

PROJECT OBJECTIVES

The primary objectives of the project were to evaluate the effectiveness of the MIEX® process and to demonstrate its applicability for treatment of raw water with differing water quality. Core objectives for bench and pilot scale testing of the MIEX® process were to determine: (1) its effectiveness in removing natural organic matter (NOM) as measured by UV254 and dissolved organic carbon (DOC); (2) its capability to reduce the formation of trihalomethanes (THMs) and haloacetic acids (HAAs); (3), its effect on molecular weight size fractions of DOC and on hydrophilic/hydrophobic organic carbon distribution (4) its effect on coagulant demand, the removal of bromide, and the formation of brominated organics, as well as on chlorine demand, and ozone demand, and (5) capability to retain resin capacity through multiple cycles of regeneration. Testing was also conducted at selected sites to assess removal of sulfide, perchlorate, and arsenic, all of which are anions that can be potentially removed by the MIEX® process.

Results of the full-scale operation of the Wanneroo Groundwater Treatment Plant (WGTP) in Western Australia are combined with results from bench and pilot scale studies to develop additional guidance for planning and implementation based on actual conditions of operation.

Test Sites

Bench and pilot scale testing of the MIEX® process was conducted at the following four facilities:

- Maytum Water Treatment Plant in New Port Richey, FL, a member of Tampa Bay Water System.

- Hanahan Water Treatment Plant in Charleston, SC, operated by the Commissioners of Public Works of Charleston.
- Fort Thomas Water Treatment Plant in Fort Thomas, KY, operated by the Northern Kentucky Water District.
- Alfred Merritt Smith Water Treatment Facility in Las Vegas, NV, operated by the Southern Nevada Water Authority.

The Maytum Water Treatment Plant (MWTP) treats groundwater with aeration for sulfide removal and chlorination for disinfection, oxidation of sulfide, and reduction of color. The MIEX® process was tested at this facility for removal of total organic carbon (TOC) to control disinfection by-products to meet future regulations and for its potential to remove sulfide.

The Hanahan Water Treatment Plant (HWTP) treats surface water sources that contain high levels of TOC and bromide. Bromide is a significant factor that affects disinfection by-product formation at this location and its removal by MIEX® resin would have a substantial benefit. In addition, the effect of the MIEX® process on the downstream coagulation process was an important consideration as the removal of NOM can reduce coagulant demand.

The Fort Thomas Water Treatment Plant (FTWTP) treats Ohio River water with conventional coagulation, flocculation, sedimentation, and filtration. The water quality of the Ohio River is characterized by moderate alkalinity and TOC concentrations, and periodically high turbidities. Testing was performed during a period of high rainfall and raw water turbidity reached 150 ntu and averaged 78 ntu. The goals for the MIEX® process at this location are primarily directed towards removal of DBP precursors and enhancement of the coagulation process.

The Alfred Merritt Smith Water Treatment Facility (AMSWTF) treats water from Lake Mead in Southern Nevada and serves the Las Vegas area. The source water is characterized by low turbidity and organics content that can be treated with relatively low doses of ferric chloride as the coagulant in conjunction with a polymer for conditioning preceding direct filtration. The plant is being upgraded by incorporating a pre-ozonation step. This is a plant where opportunities for enhanced coagulation are limited by constraints of the direct filtration process. In addition, pH adjustment for enhancing organics removal by coagulation is difficult at this location because of the high alkalinity of the source water. Therefore, organics removal capabilities are limited in the absence of the MIEX® process or other processes such as activated carbon adsorption or nanofiltration. In addition, the MIEX® process was considered at AMSWTF for its potential to remove other anions such as bromide, arsenic, and perchlorate.

CHAPTER 2
LITERATURE REVIEW

The magnetized ion exchange (MIEX®) process was developed to remove natural organic matter (NOM) from drinking water. This process concept is based on the fact that much of the NOM in natural water has a negative surface charge that allows removal by anion exchange resins (Symons et al. 1995, Fu and Symons 1990, Semmens et al. 2000). Thus, anion exchange offers an alternative to NOM removal processes such as granular activated carbon (GAC) adsorption and nanofiltration (NF).

The benefits of NOM removal are widely recognized for reducing the potential for formation of disinfection by-products. NOM removal strategies have become a key component of treatment technologies outlined in the Stage 1 Disinfectants/Disinfection By-products Rule as a consequence (USEPA 1998b). Criteria for this rule incorporate enhanced coagulation for NOM removal as one component. However, many utilities have found that it may be necessary to adjust their treatment beyond enhanced coagulation to meet the THM and HAA5 limits stipulated in this rule, and may have to make further adjustments to meet the more stringent limits of the forthcoming Stage 2 Disinfectants/Disinfection By-products Rule. Therefore, technologies for removing NOM, such as the MIEX® process, could be potential components of a "toolbox" of available strategies for compliance.

The magnetized anion exchange adaptation of the ion exchange process was developed in the 1990s to accommodate the use of a mixed tank as opposed to the traditional fixed bed ion exchange system. Resin recovery is accomplished by embedding magnetic iron particles in the resin, which has resulted in recoveries of 99.9 percent or better in pilot and demonstration scale testing at sedimentation overflow rates exceeding 4 gpm/sq ft (Lange et al. 2001).

Approximately 5 to 10 percent of the recovered resin is regenerated in a batch process using an 8 to 12 percent NaCl solution, most of which can be recovered and reused. Reuse in nine or more cycles has been demonstrated, with makeup salt added to compensate for the chloride lost during the process (Lange et al. 2001). This method of conserving and applying recycle regenerant substantially reduces the quantity of spent regenerant to be disposed. At present, the process is being applied at full scale at two plants in Australia: the 112.5 ML/day (30 mgd) facility at the WGTP in Western Australia (Lange et al. 2001 and Smith et al. 2002) and a 2.5 ML/day (0.66 mgd) facility at the Mt. Pleasant Water Treatment Plant in Southern Australia (Drikas 2003).

APPLICATION OF ION EXCHANGE FOR REMOVAL OF NOM

Ion exchange is a highly selective and stable NOM removal process which has the potential to produce water of high quality (Brattebo et al. 1987). NOM removal can be achieved using anion exchange resins as a result of the inherent negative charge characteristics of much of the NOM present in natural water. Anion exchange resins are typically produced by utilizing covalent bonding to attach positively charged tertiary or quaternary amine groups to a crosslinked polymer matrix. The resin is brought into contact with a solution containing a negatively charged ion, such as chloride, which becomes electrostatically bound to the positively charged sites of the resin. The resin can be designed so that the selectivity coefficient for NOM composed of hydrophobic and hydrophilic acids is higher than that for chloride, allowing the ions to be exchanged in solution. Raw water is passed through a fixed bed of resin in a conventional application of ion exchange, and target anions are retained on the resin in exchange for chloride. The capacity of the resin for specific anions depends on the relative concentration of ions in the water and the selectivity of the resin. The resin has a finite ion exchange capacity, related to the number of exchange sites. Breakthrough occurs when the resin becomes exhausted relative to the target anions. Regeneration involves passing a concentrated brine or a caustic brine through the resin bed to reverse the process and replace the attached anions with chloride. The regeneration is facilitated by the high concentration gradient for chloride uptake that results from exposure to concentrated brine.

Adsorption of NOM by ion exchange has been found to be more successful with more hydrophilic resins (Bolto et al. 2002). Some humic substances are removed by a simple anion exchange process, and some portion of the humic material is removed by surface adsorption involving van der Waals forces, particularly in acidic media where the humic substances are not protonated (positively charged) (Fu and Symons 1990).

Weak base anion exchange resins contain tertiary amine groups that are effective only in the acidic pH range where the functional groups are protonated. Strong base anion exchange resins contain quaternary amine groups that are positively charged over a wide range of pH values and are able to remove most negatively charged organic compounds (Anderson and Maier 1979, Kim and Symons 1991, Fu and Symons 1990, Fettig 1999). At neutral to basic pH values, much of the NOM exists as negatively charged anions that should be susceptible to removal by anion exchangers. Strong base resins have been shown to be superior to weak base resins with respect to their rate of uptake, capacity, and regeneration efficiency (Boening et al. 1980, Brattebo et al. 1987, Rook and Evans, 1979).

The resin skeleton and pore structure are also important parameters for effective NOM removal. The amount of swelling of the resin is determined by the resin skeleton, and highly swelled resins with acrylic skeletons have been shown to remove more aquatic organic substances than resins with rigid, highly aromatic, polystyrene skeletons (Fu and Symons 1990). The pore structure of anion exchange resins can be classified as either gel-containing or porous. Gel-containing resins have virtually no pores, whereas macroporous resins have a true pore phase in addition to the gel phase. They therefore have internal surface area and are more accessible to organic molecules than the gel-containing resins (Fu and Symons 1990). It has been shown that, regardless of their composition, resins with high water content and open structures are the most efficient at removing NOM (Bolto et al. 2002). Other studies have shown that ion exchange is an effective method for the removal of NOM that is associated with DBP formation (Rook and Evans 1979, Brattebo et al. 1987) and that strong base resins with a polyacrylic skeleton, quaternary amine functional groups, and macroporous structures are the most effective for NOM removal.

Magnetic Ion Exchange (MIEX®)

Orica Ltd in Victoria, Australia, manufactures the strong base magnetic ion exchange resin, MIEX®, which has been formulated to remove anionic (acid) constituents of NOM, such as humic substances. This resin was developed through research by the Australian Water Quality Centre (AWQC) in collaboration with Orica and the Commonwealth Scientific and Industrial Research Organisation (CSIRO). The MIEX® resin combines traditional strong base anion exchange properties, such as quaternary amine functional groups, a polyacrylic, macroporous structure, and medium pore size and porosity, with several additional concepts, such as a particle size 2 to 5 times smaller than conventional resins (150 to 180 μm diameter) and a magnetic iron component, which makes it possible to use the resin in a continuously stirred tank, and to recover it magnetically for recycling (Slunjski et al. 2002). When the resin particles are suspended in solution and mixed, they disperse and expose their surface for reaction; however, when the agitation stops, the particles quickly form floc due to the magnetic interactions, and settle. Special properties of the resin also include an open, packed structure with a large pore volume, which makes the resin suitable for NOM removal. The resin is added to the mixing tank as a suspension in volumetrically measured concentrations, and the concentration, regeneration rate, and contact time can be altered to achieve the targeted NOM removal.

The small MIEX® resin can exhibit higher rates of exchange or adsorption kinetics because it has a larger external specific surface area and more active exchange sites on the surface of the beads. Desorption of DOC during resin regeneration is achieved by reversing the adsorption reaction. To do this the resin is treated with a highly concentrated NaCl solution. The high chloride concentration gradient in the liquid phase causes the DOC to be replaced with chloride ions. The high salt concentration causes many conventional resins to contract during regeneration, which lowers the resin's water content and makes the release of the NOM slower and more difficult to achieve (Fu and Symons 1990).

The resin has a broad range of pH tolerance that allows exposure to both acidic and basic conditions during regeneration. Regeneration under acidic conditions allows removal of metal precipitates that might cause mineral fouling; alkaline regeneration increases the solubility of large molecular weight organics to facilitate their removal (Slunjski et al. 2000). These characteristics are intended to allow extended operation of the MIEX® process with minimal capacity loss (Bourke et al. 1999).

One of the attributes of the MIEX® process is that it can be used to treat raw water directly, without pretreatment. Accordingly, the process can be readily retrofitted to existing conventional water treatment plants (Slunjski et al. 1999). In addition, the stirred contacting system requires a considerably smaller amount of hydraulic gradient than many other NOM removal technologies and the potential for requiring additional pumping is less as a consequence. The MIEX® process contributes a small amount of suspended solids, in the form of resin carryover from the separator; therefore it must be followed by some form of turbidity removal (Slunjski et al. 2000).

It is well established that, for most natural waters, the concentration of DOC tends to control the coagulant dosage required for effective turbidity removal (Edzwald et al. 1979). Since the MIEX® process removes a significant portion of the DOC, it should reduce both the coagulant demand (Singer and Bilyk 2002, Cook et al. 2001, Morran et al. 1996) and the DBP formation potential of the water, and can result in a reduction in chemical costs and costs for waste sludge handling and disposal. It can also reduce the use of caustic soda (for pH adjustment) and disinfection chemicals (Delphos et al. 2001). However, process performance may be seasonally affected. Semmens et al. (2000) found that in St. Paul, Minnesota, the effectiveness of the resin dropped by 14 percent during the winter months. It was suggested that the difference may be attributable to changes in NOM chemistry and/or temperature.

The MIEX® process has been shown to reduce DBP formation potential (Singer and Bilyk 2002) and transphilic and hydrophophilic acids (Lee et al. 2003), as well as hydrophobic acids. Removal of transphilic and hydrophilic acids is significant as neither coagulation nor activated carbon adsorption is effective in removing these fractions of NOM. Bolto et al. (2002) has also shown that coagulants and ion exchangers can remove different NOM compounds. NOM removals can therefore be significantly increased by combining coagulation and ion exchange. Coagulants typically show preference for large molecules, while resins may also remove smaller, highly charged compounds. The reduction in organic material by the combination of coagulation and magnetized anion exchange may also reduce chlorine demand and improves the stability of the chlorine residual (Cook et al. 2001), which will result in reduced chlorine use and in a more stable finished water.

The MIEX® resin can also remove inorganic negatively charged constituents from water such as bromide, sulfate, sulfide, and arsenate (Bourke et al. 2001, Singer and Bilyk 2002). Therefore, the process has a potential for broad applicability. The extent of removal of these anions depends on the competition between other anions in the source water.

Singer and Bilyk (2002) have shown that THM and HAA formation potential decreased as a result of enhanced coagulation with alum, but the decrease was even more substantial as a result of treatment with MIEX® resin. MIEX® resin can also remove bromide under the proper conditions, particularly in low alkalinity and low TDS waters, where there is little competition from bicarbonate and other anions (Singer and Bilyk 2002). Removal of bromide may decrease the potential for formation of more harmful bromine-containing DBPs.

AQUATIC NATURAL ORGANIC MATTER

All raw water sources contain organic material. The majority of the organic matter is derived from decaying vegetation and dead organisms, and is referred to as natural NOM (Collins et al. 1986; Reckhow et al. 1990; Croue et al. 1999). It is generally quantified in terms of the total organic carbon (TOC) concentration in the water. TOC can be divided into particulate and dissolved organic carbon (DOC). DOC is operationally distinguished from TOC as the portion of organic carbon that passes through a 0.45μm filter (Danielesson 1982). The term NOM generally refers to the dissolved fraction of the organic carbon generated by natural processes. It represents a complex, heterogeneous mixture of both high and low molecular weight species, primarily organic acids, with different solubilities and reactivities (Aiken and Leenheer 1993, Croue et al. 1999). The complexity of NOM has made it the subject of many studies aimed at trying to characterize and describe its chemical and reactive properties.

NOM derived from different sources has distinctive chemical characteristics associated with the source materials. For example, humic substances produced by decay of natural vegetation contain large amounts of aromatic carbon, and are high in phenolic content and low in nitrogen. NOM derived from microbes, such as algae and bacteria, contain relatively large amounts of nitrogen and less aromatic carbon and phenols than humic substances. The relative contribution of each group to the overall DOC concentration also changes with the water source. The heterogeneous nature of NOM makes it difficult to characterize (Aiken and Leenheer 1993, Croue et al. 1999), but the isolation of functionally distinct fractions of DOC is a common approach to studying NOM. NOM is often characterized in terms of its molecular weight, hydrophobic/hydrophilic content, and UV-absorbance at 254 nm.

The molecular weight characterization of NOM is most commonly determined by size exclusion chromatography (Chin et al. 1994) or membrane filtration, such as ultrafiltration. Ultrafiltration membranes can be used to separate different fractions of NOM using various molecular weight cutoffs from 500-100,000 daltons (Aiken and Leenheer 1993). Molecular size studies have found humic acids to have a larger molecular size than fulvic acids (Reckhow et al. 1990), which have a higher average molecular weight than transphilic acids (Aiken et al. 1992). Humic acids are operationally defined as the fraction of humic substances that are soluble at alkaline pH but precipitate at acidic pH, whereas fulvic acids are the fraction of humic substances that are soluble at both alkaline and acidic pH (Aiken 1985).

The relative hydrophobicity of the different fractions of NOM is significant because of the relative reactivity of each fraction. The differences in the sorption characteristics of XAD-8 and XAD-4 resins are used to isolate these fractions from raw water. The nonionic XAD-8 resin polymer was developed by Rohm and Haas for isolation of humic substances (Aiken 1985, Aiken et al. 1992, Malcolm and McCarthy 1992, Aiken and Leenheer 1993, Croue et al. 1999) and is now considered an integral part of the reference method for analysis of humic substances. After filtration through a 0.45μm membrane, the sample is acidified to pH 2 and passed through a bed of XAD-8 resin, onto which the hydrophobic acids are adsorbed. When the XAD-8 resin is followed with an XAD-4 resin, NOM can be separated into three fractions. The hydrophobic acid, or humic, fraction of NOM is adsorbed at acidic pH onto the XAD-8 resin column. Passage of the sample through the XAD-4 resin results in the separation of the transphilic acids of intermediate polarity that adsorb onto the resin; the hydrophilic fraction passes through both the XAD-8 and XAD-4 columns without adsorption. Each column is separately back-eluted with 0.1N NaOH to obtain the hydrophobic acid fraction and the transphilic acid fraction, respectively (Aiken 1985, Aiken et al. 1992, Malcolm and McCarthy 1992, Aiken and Leenheer 1993, Croue et al. 1999). Research has shown that the hydrophobic fraction generally consists of larger molecular weight (MW) acids with lower charge density, i.e., the humic and fulvic acids. The hydrophilic fraction includes the acids of lower MW and higher charge density, such as carboxylic acids and organic bases. The transphilic fraction is the fraction of DOC that is of intermediate polarity.

Organic compounds that are aromatic in structure or that have conjugated double bonds absorb ultraviolet (UV) light. The specific UV absorbance (SUVA) is a useful measurement to characterize NOM in water supplies. It is defined as the UV absorbance per unit concentration of DOC, expressed as L/mg-m. For a given water, the SUVA value of the humic acids is greater than that of the fulvic acids, which is greater than that of the transphilic acids. This is the same order as their aromaticity and molecular size (Croue et al. 1999). A SUVA value of 4 to 5 indicates that the organic fraction of the water is composed largely of aquatic humic material; the DOC is relatively hydrophobic, aromatic, and of high molecular weight. SUVA values less than 3 indicate that the DOC is composed largely of non-humic materials, and the organic matter is relatively hydrophilic, less aromatic, and of lower molecular weight (Edzwald 1993).

NOM has many detrimental effects on water treatment, including reaction with disinfectants and increasing the disinfectant demand; it is the primary precursor in the formation of disinfection by-products (DBPs) (Rook 1974; Rook 1976; Stevens et al. 1976; Oliver and Thurman 1983; Reckhow et al. 1990; Singer 1994; Singer 1999; Kargalioglu et al. 2000; Gallard and von Gunten 2002; Liang and Singer 2003). It increases the coagulant demand and consequently, sludge production (Edzwald et al. 1979, Edzwald 1993, White et al. 1997, Singer and Bilyk 2002), and binds with metals, making them more difficult to remove (Schmitt et al. 2003). It also serves as a food source for microorganisms, resulting in bacterial regrowth in distribution systems (Owen et al. 1995), and reduces the effectiveness of membrane processes by causing irreversible fouling (Farahbakhsh et al. 2003). An understanding of the properties of NOM and its effects on water treatment and finished water quality is therefore very beneficial with respect to optimization of treatment and finished water quality.

Formation of Disinfection By-Products

Disinfection by-products are formed as a consequence of reactions involving oxidizing chemicals used as disinfectants. The byproducts of reactions between these disinfectants and NOM have been an area of particular concern. The most widely recognized DBPs of reactions involving chlorine, the most widely used disinfectant, are THMs and HAAs (Christman et al. 1983; Krasner et al. 1989; Cowman and Singer 1996). There are many other by-products formed during chlorination, including, but not limited to, haloacetonitries, haloketones, cyanogens halides, oxyhalides, and aldehydes. The chlorine- and bromine- containing THM and HAA species formed during chlorination are listed in Table 2.1.

DBP formation is affected by the concentration and characteristics of NOM (Reckhow et al. 1990), as well as by raw water quality and treatment techniques. The quantity and speciation of DBPs in finished water varies with water quality and treatment conditions, such as temperature, pH, and chlorine dose, contact time, and residual concentration. It also varies with source water bromide concentration and DOC concentration (Stevens et al. 1976, Singer 1994, Croue et al. 1999).

Table 2.1
THM and HAA species

Trihalomethanes	
Chloroform	$CHCl_3$
Bromodichloromethane	$CHBrCl_2$
Dibromochloromethane	$CHBr_2Cl$
Bromoform	$CHBr_3$
Haloacetic Acids	
Monochloroacetic acid	ClAA
Dichloroacetic acid	Cl_2AA
Trichloroacetic acid	Cl_3AA
Bromodichloroacetic acid	$BrCl_2AA$
Dibromochloroacetic acid	Br_2ClAA
Bromochloroacetic acid	BrClAA
Monobromoacetic acid	BrAA
Dibromoacetic acid	Br_2AA
Tribromoacetic acid	Br_3AA

The temperature and pH of a water have been shown to directly affect the formation of DBPs. Higher temperatures increase the reactivity of NOM with chlorine and accelerate DBP reaction kinetics; in fact, chloroform concentrations at 25°C were found to have doubled in a given amount of time from those at 3°C (Stevens et al. 1976, Rook 1976). Another important parameter is pH during chlorine oxidation/disinfection. Overall HAA formation is known to decrease with increasing pH. Dihaloacetic acid formation appears to be independent of pH, but the tri-haloacetic acid species decrease with increasing pH (Reckhow and Singer 1984; Reckhow et al. 1990; Singer 1994). Conversely, the formation of THMs increases with increasing pH (Stevens et al. 1976; Rook 1976; Reckhow and Singer 1984; Reckhow et al. 1990). The effect of pH may be interpreted as a shift in the chlorination mechanisms as the degree of protonation of the reacting species changes (Reckhow et al. 1990). The aggregate concentration of all halogenated DBPs is sometimes represented as the total organic halide (TOX) concentration, and overall TOX formation for a given water has been shown to decrease with increasing pH (Reckhow and Singer 1984, Singer 1994).

The amount and distribution of DBPs are also impacted by the chlorine dose, residence time, and free chlorine residual. Higher doses and free chlorine residual concentrations favor the formation of HAAs over THMs, the formation of trihalogenated HAAs over the di- and mono-halogentated HAAs, and the formation of chlorinated THMs and HAAs over brominated and mixed bromo-chloro species (Symons et al. 1993, Singer 1994). THM and HAA formation stops when the free chlorine residual is depleted, although the formation of some DBPs continues to a limited degree as a result of hydrolysis. While THM and HAA formation increases with increasing contact time (Stevens et al. 1976, Reckhow and Singer 1984, Summers et al. 1996) and therefore continues in the distribution system as long as a free chlorine residual persists, some of the halogenated DBPs form rapidly upon chlorination, but decay in the distribution system due to hydrolysis and continuing reactions with residual chlorine. Examples are haloacetonitriles and haloketones, other products of disinfection, which have been shown to degrade at longer residence times (Reckhow and Singer 1985; Krasner et al. 1989; Zhang and Minear 2002).

Total organic carbon concentration and ultraviolet absorbance or SUVA have been reported to serve as useful surrogates of DBP precursor concentrations (Singer et al. 1981; Edzwald et al. 1985; Krasner et al. 1996; Croue et al. 1999; Gallard and von Gunten 2002). THM and HAA formation potentials have been shown to be directly proportional to the TOC concentrations (Babcock and Singer 1979; Edzwald et al. 1985; Singer 1999) and the SUVA is an indicator of the aromatic carbon content of the NOM. THMs formed per unit of DOC increase as the SUVA increases (Reckhow et al. 1990; Krasner et al. 1996; Croue et al. 1999). While all fractions of NOM react with chlorine, the hydrophobic fraction of organic material has a higher SUVA and higher HAA and THM formation potential than the corresponding hydrophilic fraction (Liang and Singer 2003). However, hydrophilic carbon plays an important role in DBP formation, especially for waters with low humic content. Aquatic humic acids have been shown to consume more chlorine, and to produce more TOX (Oliver and Thurman 1983; Reckhow et al. 1990; Croue et al. 1999), chloroform, trichloroacetic acid, dichloroacetic acid, and dichloroacetonitrile on a per unit of available organic carbon basis than aquatic fulvic acids extracted from the same source water (Reckhow et al. 1990; Croue et al. 1999). Fulvic acids, in turn, are more reactive with chlorine than transphilic acids (Croue et al. 1999). However, because fulvic acids have a higher carboxylic and phenolic acidity (and thus a higher charge density), they tend to be more difficult to chemically coagulate than humic acids, which have a lower charge density (Owen et al. 1995).

The formation of brominated DBPs increases with increasing bromide ion concentrations or with higher bromide to chlorine ratios (Singer 1999) and will be discussed below. There are also seasonal effects on the rate and extent of DBP formation. Temperature considerations, which affect reaction kinetics, account for some of this, while the nature of the organic precursors may vary seasonally, and the concentration of bromide may vary between the wet and dry season (Singer 1994).

Hydrophilic fractions react more readily with bromine than hydrophobic fractions (Liang and Singer 2003). The most significant source of bromide in raw water supplies is salt water intrusion (Krasner et al. 1994). However, the presence of bromide in water supplies is also attributable to connate water (ancient, geologically trapped seawater), oil-field brines, and industrial and agricultural chemicals. Bromide in water supplies has been found to have a great effect on both the production and speciation of DBPs. Bromide is oxidized by chlorine to hypobromous acid which subsequently forms the brominated DBP species (Singer 1994):

$$HOCl + Br^- \rightarrow HOBr + Cl^-$$
$$HOCl + HOBr + NOM \rightarrow \text{Bromo- and Chloro- DBPs}$$

Cowman and Singer (1996) have shown that the relative distribution of mono-, di-, and trihalogenated HAA species is independent of Br^- concentration, which suggests that the HAA species containing bromine are probably formed through the same chemical pathways as the chlorinated species. It has also been shown that tribromoacetic acid decomposes relatively quickly to form bromoform (Zhang and Minear 2002). The two mixed chlorobromo-species, dibromochloroacetic acid and bromodichloroacetic acid, also decompose to form dibromochloromethane and bromodichloromethane, respectively, at ambient temperatures, which indicates that three of the four typical THMs can be produced to some extent by the decomposition of tri-halogenated HAAs. This constitutes a complementary pathway for the formation of THMs in drinking water, but it appears that pH in the range of 6-9 does not cause the decomposition of these brominated HAAs (Zhang and Minear 2002).

Research has shown that reducing the organic carbon concentration prior to chlorination reduces total DBP formation (Reckhow et al. 1990; Owen et al. 1995; Black et al. 1996; Singer 1999; Singer and Bilyk 2002; Liang and Singer 2003). However, the concentration of the brominated species can increase. The ratio of the source water bromide concentration to the applied chlorine dosage affects the distribution of the DBP species. A shift to the more highly brominated forms of DBPs will therefore occur as the bromide to chlorine dosage ratio increases (Symons et al. 1993; Pourmoghaddas et al. 1993; Krasner and Means 1994). The bromide-to-chlorine ratio increases with decreasing organic carbon concentration, shifting the distribution of DBPs to more brominated forms (Krasner and Means 1994; Black et al. 1996; Cowman and Singer 1996; Liang and Singer 2003).

The bromine incorporation factor (BIF) represents the average number of moles of bromine incorporated into the DBPs per mole of DBP (Ko et al. 1996; Chang et al. 2001; Miles 2002). The BIF is calculated by dividing the sum of the molar concentration of bromine in each individual THM or HAA species by the total molar concentration of THMs or HAAs. This can be used to quantitatively describe the degree of bromination in the DBP distribution, and increases as the bromide concentration increases or as the bromide to chlorine or bromide to DOC ratio increases (Ko et al. 1996; Chang et al. 2001; Miles 2002).

REMOVAL OF NOM BY COAGULATION

Coagulation can be used to remove both the dissolved and particulate fractions of NOM. The dissolved fraction is removed by precipitation or coprecipitation (Randtke 1988). During precipitation, DOC forms a complex with the dissolved coagulant and precipitates out of solution when the solubility of the complex is exceeded. DOC can also adsorb to the surface of the metal hydroxide as it precipitates (coprecipitation). The particulate NOM is removed by charge neutralization.

Research has shown that coagulation preferentially removes humic and higher molecular weight NOM (Owen et al. 1995). It has also been shown to reduce the SUVA of the water, reflecting preferential removal of UV-absorbing substances. This reduction supports the preferential removal of the humic or hydrophobic fraction of NOM (Stevens et al. 1976; Kavanaugh 1978; Babcock and Singer 1979; Reckhow and Singer 1984; Collins et al. 1985; Edzwald 1993; Krasner and Amy 1995; Owen et al. 1995).

Enhanced coagulation is used to achieve reductions in TOC concentrations that comply with regulatory criteria developed for the Stage 1 Disinfectants/DBP Rule. The enhanced coagulation requirements mandate reductions in TOC concentrations based on source water TOC and alkalinity concentrations. The required percent removals based on alkalinity and TOC are shown in Table 2.2.

Table 2.2
Percent TOC removal requirements for enhanced coagulation

TOC Concentration (mg/L)	Alkalinity (mg/L as $CaCO_3$)		
	0-60	>60-120	>120
>2.0-4.0	35 %	25 %	15 %
>4.0-8.0	45 %	35 %	25 %
>8.0	50 %	40 %	30 %

Waters that contain large amounts of low molecular weight NOM can be difficult to treat by coagulation, especially if the waters are low in turbidity (Edzwald 1993). The smaller molecules are less effectively removed by coagulation, and therefore will require treatment using adsorbents or biological methods to yield significant removals. There are many water sources in the US where coagulants cannot achieve the required TOC removal due to the characteristics of the TOC. Waters with high alkalinity and low to moderate TOC concentrations with low SUVA values pose a serious challenge for TOC removal by coagulation (White et al. 1997).

CHAPTER 3
EXPERIENCE WITH FULL SCALE APPLICATION OF THE MIEX® PROCESS AT THE WANNEROO GROUNDWATER TREATMENT PLANT IN WESTERN AUSTRALIA

The Wanneroo Groundwater Treatment Plant (WGTP), which supplies drinking water for the City of Perth, Western Australia has been using the MIEX® process since 2001. The WGTP treats groundwater containing high concentrations of TOC, iron, and non-sulfide-reduced-sulfur (NSRS). The MIEX® portion of the plant treats half of the 225 ML/d (60 mgd) total plant flow. (Lange 2001)

A much smaller MIEX® facility was also installed at the Mt. Pleasant Water Treatment Plant in the Adelaide Hills area of South Australia. This facility treats 2.5 ML/d (0.66 mgd) of water from the Murray River. The MIEX® process has been in operation at this plant since July 2001 (Drikas. 2003).

DESCRIPTION OF WGTP FACILITIES AND OPERATION

MIEX® facilities were installed at the WGTP as a 112.5 ML/d (30 mgd) retrofit following initial aeration and iron oxidation in conjunction with the addition of chlorine. The facility consists of two mixed contactors followed by six hopper-bottomed sedimentation tanks. The regeneration system includes two resin regeneration tanks, one regenerant storage tank, and two salt saturating systems for preparation of the regenerant brine. Effluent from the process is pumped to a solids contact clarification/granular media filtration system that uses alum for coagulation.

The results of pilot plant pre-design studies indicated that the contact time for a given level of performance was related to resin concentration. For example, the targeted DOC removal could be achieved either at a contact time of 10 minutes and a resin dose of 20 mL/L or at a contact time of 30 minutes and a resin dose of 8 mL/L. Significantly, there was little difference in the rate of resin loss at these two concentrations. As the system could be operated at a higher resin concentration without increased resin loss, a design based on the 10-minute contact time was determined to be more economical, and was therefore selected for implementation. The cost of the regeneration system under this approach was determined to be somewhat higher because of the need to handle a higher resin concentration, but this was not enough of a cost impact to offset the cost reduction for the contactors.

Based on the pilot studies, a sedimentation hydraulic loading rate of 15 m/hour (6.1 gpm/sq ft) was selected for design of the full-scale facility in order to provide for capture of 99.98 percent of the MIEX® resin. The sedimentation tanks were designed with Fitch-type feedwells to promote particle-to-particle contact and subsequent agglomeration of the resin under conditions conducive to formation of a settleable floc that would yield efficient retention of the resin. The Fitch-type feedwells are vertically mounted cylinders that extend to just above the hopper bottom of the tank where the solids accumulate. Flow is introduced tangentially in opposing directions through two influent pipes (Smith et al. 2002).

Operating experience for the full-scale installation has demonstrated that a resin capture of 99.98 percent is achievable. Initial evaluations indicated a resin loss rate of 4.2 liters per million liters (L/ML) of water treated. Subsequent evaluations showed a loss rate of 3.8 L/ML treated (Smith. 2003). Levels of loss have more recently been reduced to 3.0 L/ML (Botica 2004)

The two regeneration basins operate in a batch mode, with one basin operating while the other is being filled with resin. Initial draining for these operations is through decant valves on the side of the basins which facilitate relatively rapid draining to the bottom of the sidewall, followed by a slower final draining using a vacuum pump to withdraw water from the bottom of the basin through a filter that separates the resin from the liquid. (Smith et al. 2002)

Spent resin is normally regenerated using an approximately 12 percent solution of sodium chloride to which caustic soda is added to adjust to a pH of 10. The resin to be regenerated is withdrawn from the recycle loop that transports recovered resin from the bottom of the sedimentation tanks to the MIEX® contact basin influents, and is allowed to accumulate in one of the regeneration tanks until the regeneration process is initiated. Mixing is provided to maintain the resin in suspension during regeneration. After completion of regeneration, the brine is returned to the regenerant storage tank for re-use, and the regenerated resin is conveyed to the fresh resin storage tank. From the storage tank, the freshly regenerated resin is pumped to the MIEX® process influent to be combined with settled resin recycled from the bottom of the sedimentation tanks. The percentage of resin to be regenerated is established by the ratio of the flow component of settled resin that is diverted for regeneration to the total flow of the settled resin stream that is collected from the bottom of the sedimentation tanks (Smith et al. 2002).

One design factor to be considered is whether a rinse step is to be applied to remove excess brine after completion of regeneration. The rationale is that unless the resin is rinsed, it will carry a limited amount of brine into the system, which will increase the sodium and chloride concentrations in the process effluent. However, adding the rinse step would extend the duration of the regeneration process, and thus the time cycle for alternately filling and regenerating resin in the regeneration tanks, which would ultimately lead to the need for additional regeneration basin volume and a larger volume of wastewater to be disposed. In the case of the WGTP, it was decided not to include the rinse cycle in the initial design.

The brine can be reused several times before the concentrations of DOC and other constituents increase to unacceptable levels. The chloride ion lost in each regeneration is replenished by adding sodium chloride to the brine, using one of the brine saturators designated as the "dirty" brine saturator. The second saturator is used to make up new brine solution to replace spent brine that can no longer be reused.

Another consideration in configuring the regeneration system was whether to purchase the salt in dry form and to prepare the brine on-site, or to purchase the salt in saturated liquid form. Evaluations for this location indicated that it would be more economical to purchase dry salt and saturators, and to prepare the brine on-site.

The plant also includes provisions for adding either caustic soda or hydrochloric acid to the regenerant brine. Normally, caustic soda is added to elevate the pH of the brine to 10 to 11, which enhances the release of NOM. Provisions for adding acid make it possible to (1) adjust the pH of the regenerated resin stream before it is returned to the process influent and (2) adjust the pH of the waste regenerant before it is disposed (Lange 2001). Provisions for acidifying the resin were also incorporated to reduce fouling by precipitating iron that results from the placement of an iron oxidation step immediately upstream from the MIEX® process.

The spent brine is disposed in batches and replaced with fresh brine produced by the saturators. An alternative would be to provide a continuous flow of fresh brine, while continuously removing a corresponding amount of the spent brine.

Disposal of waste brine is a critical consideration in establishing the viability of the process. At the WGTP, the brine is hauled to an approved ocean outfall. Toxicity studies performed to support this method of disposal indicated that the waste stream would be diluted up to 1,000 times in the outfall pipe (Smith et al. 2002). The volume of brine to be disposed is approximately 350 to 400 L/ML treated (350 to 400 gallons per million gallons treated) (O'Brien 2004); therefore, brine volumes to be disposed are relatively small as compared with the flow that is treated.

Resin Handling Considerations

Special provisions are required for minimizing mechanical abrasion that will accelerate the rate of resin attrition and for handling the resin slurry in the recycle and regeneration streams. These considerations affected the design of the WGTP as follows (Smith et al. 2002):

- Use of low shear hydrofoil mixers operated at 30 to 50 rpm in the contacting basins, and low shear pumps operated at 600 rpm.
- Construction of the hoppers in the sedimentation tanks at a 70° angle to facilitate collection of the resin without the need for mechanical scrapers or other collection devices that would result in physical abrasion or plugging by the accumulated slurry. Special construction techniques were required for concrete structures built at such a steep angle.
- Providing a flushing water system to keep the resin in suspension and to avoid plugging of the resin conveyance piping to remove slurry from piping segments before they are removed from service or when or a pump is taken off-line. Long radius bends are used in the piping layout to minimize attrition and the tendency for plugging.

WGTP PERFORMANCE

Raw water supply for the WGTP contains dissolved organic carbon (DOC) at concentrations up to 18 mg/L. The groundwater also contains NSRS that can produce a "swampy" taste and odor in the distribution system as a result of formation of dimethyl trisulfide. Important goals of the MIEX® process are to reduce taste and odor events and the formation of disinfection by-products.

The MIEX® process is preceded by aeration in conjunction with the addition of chlorine for oxidation of iron. Following contacting with MIEX® resin in the mixed contactors and the sedimentation tanks, the water is pumped to solids contact clarifiers where alum is added. The final particle removal takes place in granular media filters. Pilot testing demonstrated this process sequence to be capable of reducing the DOC concentration through the MIEX® treatment step by 60 percent, which when followed by alum treatment could result in DOC concentrations as low as 1.5 to 2.0 mg/L from raw water levels in the range of 12 to 14 mg/L. The SUVA for the raw water was in the range of 5.5 to 6.0. (Smith et al. 2002)

Use of the MIEX® process at the WGTP faces the potential challenge of resin fouling by precipitation of iron from the upstream preoxidation process. Despite this challenge, the application has been successful; resin fouling has been limited by controlling the pre-chlorine dose and the chlorine residual through the upstream oxidation step to a minimum level for reliably achieving goals for iron oxidation (Masters 2003).

Reducing the DOC concentration by 50 percent was an operational objective for the full-scale MIEX® process that has been generally achieved at full-scale when a regeneration rate of 5.2 percent of the recycled flow or greater is applied. Removals are slightly lower when the regeneration rate is lower than 5.2 percent.

Another factor that affects the DOC removal is the loading to the plant: the removal rate tends to be lower at higher DOC loadings. The DOC loading is affected by both the DOC concentration and the volume of water treated. It would be possible to compensate for the higher loading with a higher resin concentration, but a higher concentration will mean a higher total resin inventory, and thus a larger regeneration capacity to maintain a given regeneration percentage. This modification has not proved necessary to meet treatment goals (Masters 2003).

The performance issues associated with the full-scale process were evaluated over a period from January 20th to June 8th, 2003 (Masters 2003). Flow to the MIEX® process was varied from 35 to 100 ML per day (9.2 to 26.4 mgd), and averaged 70 ML/day (18.5 mgd). As the DOC concentration in the raw water varied from 5 mg/L to 14 mg/L, there was a significant variation in DOC loading to the MIEX® process.

DOC removal during this period averaged 48 percent, even though regeneration conditions periodically caused the regeneration rate to drop from the design value of 5.2 percent to only 3.1 percent. The regeneration system is typically operated at a fixed rate, so that the actual regeneration percentage at any given time varies with the resin concentration and the plant flow rate. Therefore, fluctuations in resin capacity and variations in the plant flow rate can cause significant variations in regeneration percentage. The fixed regeneration rate affects the efficiency of DOC removal as DOC loading is increased. Evaluations indicated that DOC removals higher than 50 percent could be achieved as long as the loadings were not excessive under this condition. DOC removal was also affected by the condition of the regenerant brine; the number of uses is based on operational decisions that depend on the nature and quantity of NOM that is present.

Resin performance during the testing period appeared to improve when a significant quantity of virgin resin was added, and tended to approach a steady-state condition after a period of acclimation.

The capability of the process to remove bromide was evaluated because of the effect of this constituent on the formation of THMs and HAAs. While bromide removal has not been a major objective for the WGTP, it was of interest as a possible added benefit of the MIEX® process. Bromide concentrations during the evaluation ranged from 0.2 to 0.5 mg/L. Removals achieved were less than 5 percent under most conditions. The salt used for constituting the brine at WGTP is derived from a solar salt that is mined rather than from a sea salt source. The extent to which it contains bromide as an impurity is unknown, but this may be one factor that affects the bromide removal.

The removal of NOM by the MIEX® process has a significant effect on the alum dose required for coagulation downstream from the MIEX® process. Side-by-side comparisons of treatment trains with and without MIEX® treatment indicated that the alum dose required for the MIEX® treated stream was approximately one-third of the dose required for the stream that was not treated with the MIEX® process. This observation was consistent with results from jar tests performed during the pilot study preceding the design of the full-scale system. Other observations include reduced chlorine demand and disinfection by-product formation (Smith et al. 2002).

The Mt. Pleasant Water Treatment Plant (MPWTP) is another location that provides full-scale operating experience under somewhat different conditions of operation, which include a design contact time of 20 minutes rather than of 10 minutes used at the WGTP. The MPWTP treats a surface water with a 5-year average turbidity of 55 ntu and a peak turbidity above 200 ntu. DOC concentrations from plant startup in November 2001 until July 2003 have ranged from 3.6 mg/L to 10.1 mg/L. The MIEX® process is applied to the entire plant flow, preceding conventional flocculation, sedimentation, and granular media filtration on one side of the plant, and as pretreatment ahead of submerged membrane filtration on the other side. With resin concentrations of 5 to 15 mL/L, DOC removals by the MIEX® process are reported to be in the range of 40 to 50 percent, and alum doses in the downstream coagulation process have been reduced. Overall DOC reductions through the plant by the MIEX® process in combination with alum coagulation have increased by 20 to 30 percent as compared to removals achieved with alum alone, chlorine demand has been reduced by 20 percent, and THM formation has been reduced by 40 percent (Drikas 2003).

COSTS

The capital costs of incorporating the MIEX® process as a retrofit to the existing WGTP plant were approximately $15 million Australian ($11.1 million US). Operating costs are summarized as follows:

Resin costs (depending on resin loss)	$50-65/ML (AUS)	$141-183/MG (US)
Non-resin operational costs:		
Hydrochloric acid	$1-1.5/ML (AUS)	$3-4/MG (US)
Salt	$5-7/ML (AUS)	$14-20/MG (US)
Caustic soda	$1-1.25/ML (AUS)	$3-4/MG (US)
Brine disposal	$3-6/ML (AUS)	$8-17/MG (US)

Weekly manpower requirement has been about 20 man-hours, with a greater requirement initially during start-up (O'Brien 2004).

CHAPTER 4
MATERIALS AND METHODS

TESTING PROGRAM

Testing included one-week of pilot plant startup/bench scale testing followed by four weeks of pilot operation at each location. All testing, as well as basic analytical procedures, were performed on-site. Specialized analyses for THMs, HAAs, DOC, molecular weight fractionation, hydrophobic/hydrophilic fractionation were performed at the University of North Carolina (UNC) at Chapel Hill. Additional analyses for constituents such as bromide, bromate, perchlorate, and arsenic were performed by each utility's certified laboratory.

Bench-scale Testing

Two sets of bench scale tests were performed at each location. The first set was performed during the week of pilot plant startup at the beginning of testing and the second set of tests was performed during the last week of pilot plant operation. Resin used in this testing was obtained from the pilot plant after a period of operation to reach steady state conditions. This allowed the use of resin that had acclimated to an actual operational state in which resin is recycled and a portion of its capacity is consumed as opposed to relying on results from virgin resin.

Bench scale testing to develop basic dose-response and kinetic information was performed according to the following sequence using standard jar test apparatus. Resin dose-response trends were developed using detention times of 7.5, 15, and 20 minutes and concentrations of 0, 5, 10, 15, and 20 mL/L of resin. UV254 and DOC dose response relationships were developed for all locations, while relations for bromide, sulfide, arsenic, and perchlorate were developed at selected locations. In the testing for arsenic and perchlorate, which was performed only at the AMSWTF, these constituents were spiked to the raw water sample and measured after testing. All other tests were performed with ambient levels of the constituents of interest

All tests were performed using 95 percent spent resin taken from the pilot plant settler underflow and 5 percent freshly regenerated resin to simulate a condition in which fresh and spent resin are blended as is the case in an actual flow-through system. Supernatant was separated from the resin at the end of the contact period and samples were withdrawn for analysis. In addition to the constituents of primary interest, analyses were performed for temperature, hardness, alkalinity, and turbidity in raw water, and pH was measured in both raw and treated water.

Coagulation studies were performed at the Hanahan WTP and the Fort Thomas WTP using a standard jar testing apparatus to develop profiles indicating the effects of the MIEX® process on coagulation. Coagulation conditions selected for both raw water and MIEX® treated water was used for THM and HAA formation studies and molecular weight fractionation studies, which were performed at the University of North Carolina. The Maytum WTP does not practice coagulation; therefore no coagulation studies were performed at this location. The full-scale process at AMSWTF involves the use of ferric chloride in a direct filtration treatment sequence that is not well replicated by conventional jar testing. Therefore, treated water from the MIEX® pilot plant was pumped to a 6 gpm ozonation, coagulation, flocculation, and direct filtration pilot plant to assess the effects of the MIEX® process as a pretreatment step for this treatment sequence.

The coagulation profiles at the Hanahan WTP and the Fort Thomas WTP were performed following the bench-scale dose response tests using the following procedures:

Step 1. Simulations of a MIEX® - coagulation process sequence were compared with a control condition that consisted of coagulated water with no MIEX® pretreatment. The MIEX® dose used for simulating the MIEX® - coagulation process sequence was selected based on observations from the dose response tests. A test matrix was developed for each set of tests in which both pH and coagulant dose were varied to produce a profile of the composite effect of these parameters on settled turbidity and UV254 (surrogate for assessing organic removal). Mixing conditions for the jar tests were as follows: 30 seconds at 100 rpm, 10 minutes each at 30 rpm, 20 rpm, and 12 rpm, followed by settling for 5 minutes. Samples were withdrawn at the end of each test to determine settled UV254, pH, and turbidity in each test jar using on-site analytical equipment. On-site raw water analyses were performed for temperature, pH, alkalinity, hardness, UV254 and turbidity. Results of coagulant dose/pH profiles of the test sequences were compiled to select the best coagulation condition for each treatment sequence based on capability to reduce UV254 in regions where a stable floc could be formed as indicated by settled water turbidity

Step 2. Coagulant conditions selected from results in Step 1 were used to assess the effect of a MIEX® –coagulation sequence on THM/HAA formation under uniform formation conditions (UFC) as compared with the control condition with coagulation only. Supernatant samples were withdrawn at the end of each test sequence to determine chlorine demand based on a 1-hour demand test. This information was used as the basis for selecting the chlorine dose to be applied for longer term chlorine demand and UFC THM/HAA formation testing. UFC THM/HAA and chlorine samples were withdrawn for analysis at intervals of 2 hours, 1 day, and 3 days. Temperature and pH of these samples were also measured.

This step of testing was also used to provide samples for molecular weight and hydrophilic/hydrophobic fractionations to compare the effects different treatment sequences on these NOM attributes. This testing was performed for the control condition with coagulation, for removal by MIEX® resin prior to the addition of the coagulant, and for the combined MIEX®–coagulation treatment sequence.

Pilot Testing

Pilot testing was performed with a specially designed 10 gpm pilot plant. This testing was performed to determine the conditions of steady state operation of the process together with regeneration attributes and capability to sustain the capacity of the resin through multiple regeneration cycles. Specific evaluations are as follows:

Step 1. During operation in the continuous flow mode, the resin inventory in the pilot plant cycles through regeneration at an average rate of once every 2.5 hours at regeneration ratio of 5 percent. Therefore, a steady-state resin condition starts to be approached in less than a day. The effects of varying resin dose and regeneration frequency were assessed based on UV254 and DOC results. Raw and treated water samples were collected daily for analyses of UV254, pH, temperature, and turbidity, and for bromide and sulfide at selected locations. Samples for DOC, alkalinity, hardness, chloride, TDS and sulfate were analyzed twice per week, when applicable. Twice weekly sampling was also performed for perchlorate and arsenic at the AMSWTF. An on-site analysis of the chloride concentration in the brine was performed before each regeneration sequence at the HWTP, the FTWTP and the AMSWTF.

Step 2. Effluent from the pilot plant was withdrawn for bench-scale testing using the test procedures described in *Bench-Scale Testing*.

Step 3. Capability to limit loss of resin exchange capacity over multiple recovery/regeneration cycles was assessed based on bench-scale testing of UV254 and DOC uptake from regenerated resin. The resin was extracted at the beginning of the bench studies and the end of the pilot operation. This resin was stored through the pilot test and the resin samples were tested in side-by-side dose-response characterizations using raw water from the respective test locations to identify any trends in resin capacity.

Ozone demand studies were performed using an existing 6 gpm ozone pilot facility at the AMSWTF. The effluent from the MIEX® pilot plant was fed to the ozone pilot plant. The ozone contactor is made-up of 12 cells that provide approximately 24 minutes of contact time at the design flow rate of 6 gpm. The following passage describes the pilot plant.

> Ozone can be added through porous diffusers mounted horizontally near the bottom of cells 1, 2, and 5. The rapid mix, flocculation, and filtration steps are scaled to be similar to the full-scale plant. There are two 10-inch stainless steel dual media filters with 20 inches of anthracite and 11 inches of sand, which is identical to the full-scale filter media. The pilot plant is equipped with on-line meters for turbidity, particle counts, pH, temperature, dissolved oxygen, conductivity, dissolved ozone, feed gas ozone concentration, and offgas ozone concentration. Each of the on-line meters is tied into computer software that displays and logs the data (Zegers et al. 2001).

MIEX® PILOT PLANT DESCRIPTION

The MIEX® pilot unit, shown schematically in Figure 4.1, was designed for an influent flow rate of 5 to 10 gpm and a contactor resin concentration of up to 20 mL/L. Regeneration of the resin can be accomplished automatically or manually. The operating cycle of the pilot plant was designed to maintain a predetermined resin concentration in the contactors while the spent resin is being regenerated in batches.

The raw water feed line is equipped with a basket strainer and a sample tap. Water can be delivered to the pilot through either a centrifugal pump or through a bypass circuit around the pump that allows this mode of operation when a pressurized source of feed water is available. The raw water enters the pilot skid and flows into the contactors. There are two 75-gallon contactors in series, each equipped with variable speed, low shear mixers and a dedicated sample port. Resin is introduced into the raw water feed line just upstream of the contactors. The pilot unit is plumbed with a manual bypass circuit that allows the first contactor to be bypassed.

Effluent from the contactors flows into the settling tank, where the resin settles and the treated water exits through a discharge line that is equipped with sample tap. The settled resin is either returned to the feed point ahead of the contactor or discharged to the regeneration tank. There are three variable speed peristaltic pumps that operate to move the resin. One pump controls the flow of resin from the fresh resin tank into the contactor, while a second pump controls the flow of resin from the settler to the regeneration tank, and the third pump controls the flow of resin from the settler to the contactor. A calibration column was provided to determine the flow rates through the peristaltic pumps.

Once the regeneration tank has filled, a regeneration sequence is initiated. The regeneration sequence can be run automatically or manually. The regeneration sequence is designed to control concentrations of resin, brine and fresh water in the regeneration tank. The resin removed from the recycle line will accumulate in a tank during the regeneration cycle to avoid contaminating the clean resin. The automated regeneration is controlled by a timer system, which can be adjusted on the touch screen.

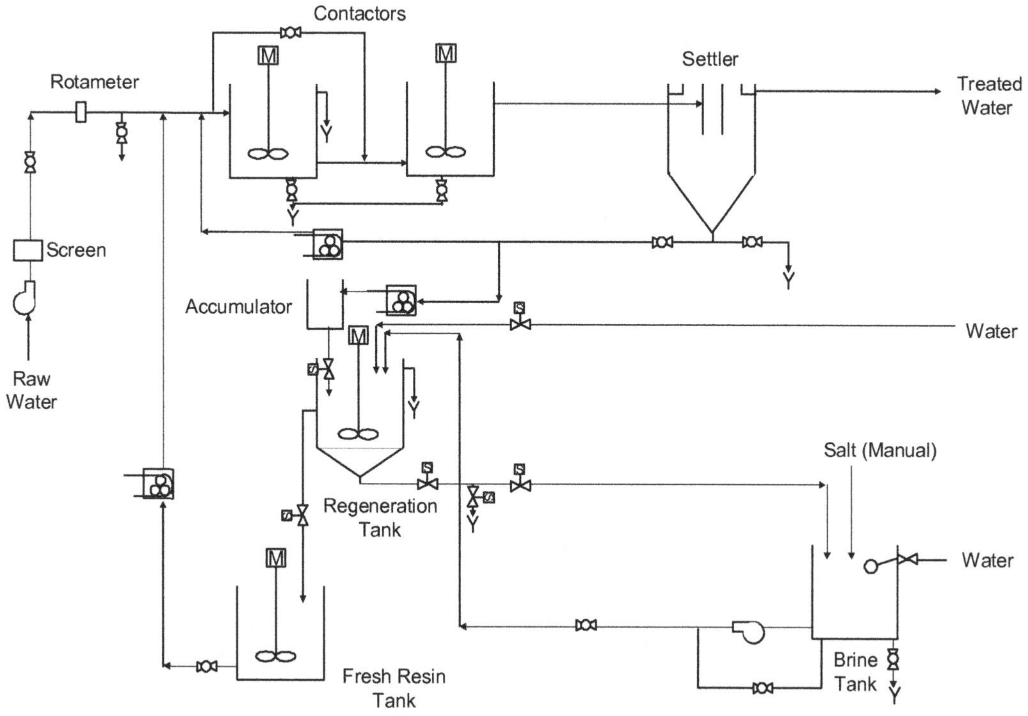

Figure 4.1 Schematic of 10 gpm MIEX® pilot plant

ANALYTICAL METHODS

Table 4.1 summarizes the analytical methods used during the study for certain basic water quality parameters. The special analyses such as DOC, bromide, THM, HAA, and XAD analytical descriptions are summarized in the text below.

Table 4.1
Analytical methods

Water quality parameter	Standard method (SM) or EPA method (EPA)	Minimum reportable limit (MRL)
pH	SM 4500H+	NA
Alkalinity	SM 2320	1.0 mg/L
Hardness	SM 2340	0.5 mg/L
Turbidity	SM 2130	NA
Temperature	SM 2550	NA
Sulfide	EPA 376.2	2.0 µg/L
Conductivity	SM 2510B	0.5 µmho/cm
Total dissolved solids (TDS)	SM 2540C	10 mg/L
Total suspended solids (TSS)	SM 2540D	10 mg/L
Dissolved ozone	SM 4500	0.01 mg/L
Chloride* (HWTP and FTWTP)	EPA 300.1A	2.0 mg/L
Chloride (AMSWTF)	EPA 300.0	2.0 mg/L
Sulfate (MWTP, HWTP and FTWTP)	EPA 300.1A	2.0 mg/L
Sulfate (AMSWTF)	EPA 300.0	2.0 mg/L
Bromate (AMSWTF)	EPA 300.1B and 317.1	5 µg/L
Perchlorate	EPA 314	4 µg/L
Arsenic	EPA 200.8	4 µg/L

* HACH silver nitrate titration method was used to determine the chloride concentration at the MWTP.

Ultraviolet (UV) Absorbance

UV absorbance measurements were made in accordance the method described by Karanfil et al. (2003). The samples were prepared by filtering through 0.45 µm Supor-450 Pall membrane filters (Pall Corporation, Ann Arbor, MI), which had been pre-rinsed with 500 mL of laboratory grade water (LGW). Measurements were made using a 1-cm quartz cell, which had been rinsed once with LGW and once with sample water prior to measurement. A minimum reporting level for this analysis of 0.010 cm^{-1} was used for this analysis.

Dissolved Organic Carbon (DOC)

DOC concentrations were measured with a Shimadzu 5000 Total Organic Carbon Analyzer equipped with an ASI 5000 auto-sampler (Shimadzu Corporation, Norcross, GA). Analysis for DOC content was conducted according to Standard Method 5310B (APHA, 1998), with one modification: the samples were acidified with 2N hydrochloric acid rather than with phosphoric acid. This was done because the Shimadzu analyzer is sensitive to excess phosphoric acid. The DOC samples were first filtered through 0.45 μm Supor-450 Pall membrane filters (Pall Corporation, Ann Arbor, MI), which were pre-rinsed with 500 mL of LGW and 10 mL of the sample.

Before measuring the samples, the TOC analyzer was calibrated using four standard solutions prepared from a 1000 mg/L organic carbon stock solution. The stock solution was made by dissolving the required amount of potassium hydrogen phthalate (Fisher Scientific, Fair Lawn, NJ) in LGW. Ultra-high purity air supplied by Holox Company (Morrisville, NC) was the carrier gas. Calibration standards were prepared fresh on the day of each analysis and a new stock solution was made every three months. The concentration of standard solutions was dependent on the suspected range of values.

After each set of ten samples, a sample of LGW and one of the calibration standards were analyzed for quality assurance purposes. All samples were analyzed in duplicate and the results were averaged to derive the reported values. If the results were not within 10 percent of each other, the samples were re-analyzed. If more than 35 samples were to be analyzed, a recalibration was programmed into the instrument using the same calibration standards for the second set of samples. A complete set of the calibration standards were also analyzed as unknown samples at the end of all analyses to monitor the possibility of sample degradation or instrumentation drift while the samples were waiting to be analyzed. A minimum reporting level of 0.5 mg/L was used for this parameter.

Bromide

Bromide samples were analyzed at the certified laboratories at the HWTP in Hanahan, South Carolina; the FTWTP in Fort Thomas, Kentucky; and the AMSWTF in Boulder City, Nevada. The procedure used at all locations was U.S.EPA Method 300.1A (U.S.EPA, 1997). A filtered 10 μL sample was injected into a 9.0 mM sodium carbonate stream and passed through a series of ion exchanger columns. The anions are separated in the eluent stream, detected by conductivity, and identified based on their retention times relative to known standards. The method specifies the use of a high capacity IonPac AS9-HC anion exchange column (Dionex Corp., Sunnyville, CA) was used with suppressed conductivity detection by direct injection (Jackson 2000). The reporting level was 0.05 mg/L at the HWTP and 0.01 mg/L at the FTWTP and the AMSWTF.

XAD Fractionation

Amberlite XAD-8 and XAD-4 resins (Rohm and Haas, Philadelphia, PA) were used to fractionate the NOM in both raw and treated waters, in accordance with the method of Thurman and Malcolm (1981). The resins were rinsed with LGW to remove any residual methanol left by the soxhelet cleaning, individually placed in a beaker of LGW, swirled, allowed to settle, and then decanted fifteen times. The resins were then loaded into two 20 mL glass columns with a final bed volume of 12 mL each. After the resin was loaded, it was flushed with 1 Liter of LGW at 4 mL/min through the top of the column to insure that all methanol had been removed and to check resin packing. Any residual organic materials in the resin were rinsed from the column by running 0.1N NaOH followed by 0.1N HCl three times through the top of the column at 4 mL/min for 10 minutes each. The two columns were connected in series with the XAD-4 column following the XAD-8 column.

All samples were filtered through pre-rinsed 0.45 μm Supor-450 Pall membrane filters (Pall Corporation, Ann Arbor, MI), which were pre-rinsed with 500 mL of LGW and 10 mL of the sample. The sample was acidified to a pH below 2.0 with concentrated HCl so that the acidic functional groups on the NOM would be protonated, thereby enhancing sorption onto the XAD resin. One liter of the sample was passed through the XAD-8 column by pumping the sample through the top of the column at 4 mL/min. The first bed volume (12 mL) of effluent was discarded. The rest of the sample was loaded onto the resin until the water level was just above the top of the resin bed, so that the resin did not dry out. The XAD-8 resin retains the hydrophobic acid fraction, allowing the transphilic acids and hydrophilic material to pass through. The XAD-8 effluent was collected and then pumped through the top of the XAD-4 column at 4 mL/min. The first bed volume (12 mL) of effluent was discarded. The sample passed through the resin until the water level was just above the top of the resin bed.

Both resin beds were back-eluted by pumping 100 mL of 0.1N NaOH through the bottom of the column at 2 mL/min to elute the materials retained on the resin. The resin was then re-acidified by pumping 0.1N HCl through the resin bed until the effluent was acidic.

The DOC concentration of the feed water, XAD-8 effluent and eluent, and XAD-4 effluent and eluent were analyzed. The measured DOC concentrations were used to determine the fraction of hydrophobic, hydrophilic, and transphilic carbon in the water samples. UV measurements of each fraction were also taken.

Ultrafiltration (UF) Fractionation

Ultrafiltration membranes (Millipore Corporation, Bedford, MA) with molecular weight cutoffs of 1,000, 10,000, and 30,000 nominal molecular weight (NMW) were received with pretreatments of glycerin to prevent drying and sodium azide as a preservative. The membranes were cleaned by suspending them in a beaker of NaCl solution for 30 minutes to remove any UV-absorbing material, followed by suspending them three times in LGW for 20 min.

Before the sample was processed, the Millipore ultrafiltration filter (Millipore Corporation, Bedford, MA) was operated at 55 psi, and 70 mL of LGW was filtered through each membrane. Seventy milliliters of each sample was filtered through each of the three membranes, and the effluent from each of the membranes was analyzed for DOC and UV254 to determine the molecular weight fractions of each water sample.

Chlorination Under Uniform Formation Conditions (UFC)

A four to six percent solution of sodium hypochlorite (Fisher Scientific, Fair Lawn, NJ) was stored in the dark at 4°C in a bottle wrapped in aluminum foil. The actual concentration of the hypochlorite solution was determined on the day of each experiment by titration with a sodium thiosulfate standard (Fisher Scientific, Fair Lawn, NJ) in accordance with Standard Method 4500-Cl B (APHA et al. 1998).

Each water sample was subjected to preliminary chlorination to determine its 24-hour chlorine demand. The UFC chlorination procedure was based on methods suggested by Summers et al. (1996). Samples were chlorinated at pH 8.0 with sufficient chlorine to produce a free chlorine residual of 1.0 mg/L Cl_2 after 24 hours storage at 20°C. A 100 mg/L chlorine dosing solution was prepared based on the strength of the chlorine stock solution, as determined by the sodium thiosulfate titration described above. A pH 6.7 borate buffer was also added to the dosing solution. A series of 100 mL sample of each water was chlorinated in 100 mL volumetric flasks using a range of doses (1-2 mg Cl_2 per mg DOC). A pH 8.0 borate buffer was added to each sample. After incubating the samples in the dark for 24 hours at 20°C, the chlorine residual of each sample was measured using a pocket colorimeter (Hach Company, Loveland, CO). The chlorine dose that produced a free chlorine residual of 1.0 mg/L after the 24 hour incubation, plus an additional 1.0 mg/L of chlorine was used for subsequent chlorination to ensure a chlorine residual after 3 days.

Each water sample was chlorinated in three 300 mL standard BOD bottles with borate buffer to ensure that the pH would remain at 8.0. The bottles were sealed headspace free and stored in the dark at 20°C. One of the bottles of each water was removed from the incubator after 2 hours, the second after 24 hours, and the third after 3 days. After measuring their free chlorine residual and pH, the samples were transferred to duplicate sets of 40-mL screw-cap glass vials with PFTE-lined silicone septa for analysis of THMs and HAAs. To reduce volatilization, the samples were poured down the side of the vials. Both sets of vials contained approximately 20 mg of ammonium sulfate (Mallinckrodt, Paris, KY) to quench the residual chlorine. In the vials to be analyzed for HAAs, 20 μL of 10 mg/L sodium azide (Sigma Aldrich Chemical, Milwaukee, WI) was added as a biocide. The vials to be analyzed for THM contained 0.7 g of phosphate buffer to standardize the pH of the samples between 4.8 and 5.5. The quenched samples were stored headspace-free at 4°C until analysis. Glassware used for chlorinating the samples was treated to be free of chlorine demand as described above.

Trihalomethanes

The UFC chlorinated samples were analyzed for all four THM species (THM4): $CHCl_3$, $CHBrCl_2$, $CHBr_2Cl$, and $CHBr_3$ using a liquid-liquid extraction gas chromatography (GC) procedure modified after Standard Method 6232B (APHA et al. 1998) and the U.S.EPA Method 551.1 (U.S.EPA 1995). Each sample was warmed to room temperature and a 20 mL aliquot was transferred to a 40 mL screw-cap glass vial with a PTFE-lined silicone septum (Laboratory Supply Distributors Corp., Mt. Laurel, NJ). Four mL of MtBE ($(CH_3)_3COCH_3$) (Sigma Aldrich, Milwaukee, WI) containing approximately 100 μg/L 1,2-dibromopropane (Sigma Aldrich, Milwaukee, WI) was added to the 20 mL sample. This served as the internal standard (IS), which corrected for any deviation in sample volume injection. Approximately 6 g of sodium sulfate (Mallinckrodt, Paris, KY), which was baked for 24 hours at 400°C, was added to each sample to aid in the extraction, and the sample vials were then Vortex-mixed (Type 16700 Mixer-MaxiMix I, Thermolyne, Dubuque, IA) for 1 minute. After mixing, the samples were allowed to sit for at least 5 minutes to allow separation of the ether layer and settling of any undissolved sulfate crystals to the bottom of the vial. At least 1 mL of the MtBE layer was transferred by Pasteur pipette to a 2 mL clear glass autosampler vial (Laboratory Supply Distributors, Mt. Laurel, NJ) and capped with an TFE-faced aluminum seal (Laboratory Supply Distributors, Mt. Laurel, NJ). To provide for re-analysis of the extract where necessary, the

MtBE layer was also transferred to a second autosampler vial. The extracts were stored in a freezer at -15°C until analysis.

The extracts were analyzed using a Hewlett-Packard Model 5890A gas chromatograph (Hewlett-Packard Company, Cary, NC) with an electron capture detector (ECD). The gas chromatographic conditions for THM analysis are shown in Table 4.2. A calibration curve was prepared in LGW by dilution of stock standards of each of the THM species (Supelco, Bellefonte, PA), which were stored in the freezer in amber vials with screw caps sealed with Teflon tape. The standards were monitored for degradation by comparing their response with that of the initial calibration of the standard. Fresh standards were prepared if the check standard was less than 20 percent of the original value. Matrix spike samples were also analyzed. These samples were spiked with THM levels 2 to 3 times the expected $CHCl_3$ concentration.

Table 4.2
Gas chromatographic conditions for THM analysis

Column	
Type	DB-1 column (Supelco, Bellefonte, PA)
Length	30 m
Internal diameter	0.25 mm
Film thickness	1.0 μm
Temperature sequence:	
10 min @ 35°C	
Increase to 150°C @ 10°C/min	
Increase to 250°C @ 25°C/min	
Hold @ 250°C for 11 min	
Total Time = 36.5 min	
Injector	
Injector volume	2 μL
Temperature	150°C
Detector	
Type	Electron capture
Temperature	300°C
Gases	
Carrier gas	Helium (Holox, Morrisville, NC)
Carrier flow	1.2 mL/min @ 35°C
Makeup gas	Nitrogen (Holox, Morrisville, NC)

From the chromatograms of the calibration solutions, the retention times of the four THM species and the internal standard were determined. The peaks of these five retention times were marked and integrated to yield peak areas, and a calibration curve was developed of the relative peak area (ratio of the peak area of the THM species to the peak area of the internal standard) for each THM concentration. This procedure was used to convert the relative peak area of the sample peaks to THM concentrations. A calibration sample was analyzed after each set of 10 samples as a check for peak consistency throughout the run. All samples were analyzed in duplicate and the average of the two was reported. A minimum reporting level of 2 μg/L was used for this parameter.

Haloacetic Acids

All five HAA species (HAA5) were analyzed using a micro liquid-liquid extraction procedure based on Standard Method 6251B (APHA, 1998) and U.S. EPA Method 552 (U.S.EPA 1995), which employs diazomethane derivatization and GC/ECD direct injection analysis (Brophy et al. 2000). Standards were obtained as a mixture of HAA6 in methanol, and as separate neat compounds of bromodichloroacetic acid, dibromochloroacetic acid, and tribromoacetic acid (Supelco, Bellefonte, PA). The standards were dissolved in MtBE (Sigma Aldrich, Milwaukee, WI) and a calibration curve was prepared in a manner similar to preparation of the THM calibration curve. New stock HAA solutions were prepared every three months.

Diazomethane was generated in an apparatus consisting of two 40 mL screw-cap glass vials with PTFE-lined silicone septa connected by 1.6 mm Teflon tubing. One septum contained a gas vent. The vial with the gas vent contained 5 mL of MtBE (Sigma Aldrich, Milwaukee, WI), with the Teflon tubing immersed in the MtBE. This vial was placed in a beaker full of ice. The other end of the Teflon tubing was not immersed in reagents. Its purpose was to collect the diazomethane vapor, which would pass through the tubing and into the MtBE in the other vial. Six mL of reagent 1 was placed in the 40 mL vial without the vent. Reagent 1 consisted of 3.3 g Diazald (Sigma Aldrich, Milwaukee, WI), 5 mL Carbitol (Sigma Aldrich, Milwaukee, WI), and 5 mL MtBE (Sigma Aldrich, Milwaukee, WI). 6 mL of reagent 2 was then placed in the same 40 mL vial. Reagent 2 consisted of 6 mL LGW, 10 mL high-purity methanol (Burdick and Jackson, Muskegon, MI), and 4 mL potassium hydroxide (Fisher Scientific, Fair Lawn, NJ). After adding reagent 2, the vial was quickly capped. Vapors began to form immediately and the MtBE in the other vial began to turn yellow. The vapor was collected until the MtBE turned a dark yellow. The vial was then capped with a screw-cap fitted with a PTFE-lined silicone septum and placed in an explosion-proof freezer. All glassware used in diazomethane preparation was soaked in a 5 N NaOH bath for at least 24 hours to destroy any unreacted diazald and residual diazomethane prior to cleaning.

Before extraction, the samples were allowed to come to room temperature, and then a 20 mL aliquot of each sample was transferred to a 40 mL screw-cap glass vial with a PTFE-lined silicone septum. All samples were analyzed in duplicate. 20 μL of 20 mg/L 2,3-dibromopropionic acid (surrogate) solution (Supelco, Bellafonte, PA) was added to each sample. The surrogate solution was stirred with the pipette tip to evaluate the efficiency of derivatization and extraction. Each sample then received 1.5 mL of concentrated sulfuric acid and was placed into a container of ice until its temperature dropped to room temperature. Four milliliters of MtBE ($(CH_3)_3COCH_3$) (Sigma Aldrich, Milwaukee, WI) containing approximately 100 μg/L of 1,2-dibromopropane (Sigma Aldrich, Milwaukee, WI) was added to the 20 mL sample to serve as the internal standard (IS), which corrected for any deviation in sample volume injection, followed by 10 g of prebaked sodium sulfate (Mallinckrodt, Paris, KY) to extract the non-dissociated acidic HAA species. The vial was vortex-mixed (Type 16700 Mixer MaxiMix I, Thermolyne, Dubuque, IA) for one minute and allowed to rest until the ether layer separated and the saturated sodium sulfate crystals settled to the bottom of the vial. The MtBE layer was then transferred with a Pasteur pipette to a 2 mL volumetric flask with a screw top and PTFE-lined septum and filled to the mark. Approximately 50 mg of anhydrous, powdered magnesium sulfate ($MgSO_4$) (Sigma Aldrich, Milwaukee, WI) was added to each of the 2 mL volumetric flasks. The magnesium sulfate was used as a drying agent to minimize the effects of water on the derivatization process (Brophy et al. 2000). After allowing the drying agent to settle, 225 μL of diazomethane solution was added and the flasks were refrigerated for 15 minutes. After taking the samples out of the refrigerator, they were inspected for the presence of a faint yellow color, which indicates that sufficient diazomethane was present to drive the esterification process. The samples were allowed to come to room temperature and a small rounded scoop of silicic acid powder (J.T. Baker, Phillipsburg, NJ) was added to each flask. The acid quenched the residual diazomethane and the samples became colorless. At least 1 mL of the derivatized sample was then transferred to a 2 mL auto-sampler vial (Laboratory Supply Distribution Corporation, Mt. Laurel, NJ) and capped with a TFE-faced aluminum seal (Laboratory Supply Distribution Corporation, Mt. Laurel, NJ). A set of six calibration solutions and matrix spike samples were extracted using the same procedure as used for the samples. Derivatized sample extracts and standard extracts were analyzed on a Hewlett Packard model 5890A GC with an ECD (Hewlett Packard Company, Cary, NC). The gas chromatographic conditions for the HAA analysis are summarized in Table 4.3.

Table 4.3
Gas chromatographic conditions for HAA analysis

Column		
	Type	DB-1 column (Supelco, Bellefonte, PA)
	Length	30 m
	Internal diameter	0.25 mm
	Film thickness	1.0 μm
	Temperature sequence:	
	21 min @ 37°C,	
	Increase to 136°C @ 5°C/min	
	Hold for 3 min	
	Increase to 250°C @ 20°C/min	
	Hold for 3 min	
	Total Run Time = 52.5 min	
Injector		
	Injector volume	2 μL
	Temperature	180°C
Detector		
	Type	Electron capture
	Temperature	300°C
Gases		
	Carrier gas	Helium (Holox, Morrisville, NC)
	Carrier flow	1.2 mL/min @ 37°C
	Makeup gas	Nitrogen (Holox, Morrisville, NC)

At the beginning of each analytical run, two MtBE solvent blanks and two MtBE samples containing the internal standard were injected to condition the GC and to verify that there were no interferences. From the chromatograms of the calibration solutions, the retention times of the nine HAA species, the internal standard, and the acid surrogate were determined. The peaks at these eleven retention times were marked and integrated to yield peak areas. A calibration curve of the relative peak area (ratio of the peak area of the HAA species to the peak area of the internal standard) for each HAA species was prepared. This procedure was used to convert the relative peak areas of the sample peaks to HAA concentrations. A calibration sample was run after each set of 10 samples as a check throughout the run to ensure that sample peaks were consistent. A minimum reporting level of 2 μg/L was used for this parameter.

GLASSWARE, SAMPLE BOTTLES AND REAGENT PREPARATION

Glassware used in sample collection and general laboratory use, excluding all volumetric pieces, were soaked in Alkonox powder detergent overnight, rinsed with tap water three times, and then rinsed three times with laboratory grade water (LGW; Dracor Inc., Durham, NC). The LGW was produced by passing tap water through a Dracor (Durham, NC) water treatment system. The Dracor system consisted of a 1.0 μm pre-filter followed by 0.5 ft^3 of activated carbon and two mixed-bed deionizers. The activated carbon removed chlorine and organic material while the mixed-bed deionizers served as a polishing step to remove dissolved salts. The glassware was then soaked in a 10% nitric acid bath for 12 hours, rinsed three times with LGW, rinsed three times with methanol (Mallinckrodt Baker, Phillipsburg, NJ), and then oven-dried at 180°C. Volumetric glassware was prepared in the same manner as the general glassware except that they were air-dried instead of oven-dried. Glassware used for chlorine residual measurement and chlorination (see below) was prepared in the same manner as the general glassware including soaking 12 hours in a solution of approximately 100 mg/L Cl_2 prepared from 5-7% sodium hypochlorite (Fisher Scientific, Fair Lawn, NJ) and then rinsed three times with LGW. All plastics (caps and septa) were soaked in tap water for 12 hours, rinsed with LGW three times, rinsed with methanol three times and allowed to air-dry.

DOC samples were collected in 40 mL glass screw cap vials with polytetrafluoroethylene (PFTE) lined silicone septa (Laboraty Supply Distributors Corp., Mt Laurel, NJ). The samples were shipped in coolers packed with ice packs. Cubitainers were sent in fitted cardboard containers (Hedwin Corporation, Baltimore, MD). All shipments to the University of North Carolina were provided by an overnight carrier. The samples were refrigerated at 4°C until analysis.

All reagents used were ACS grade quality.

CHAPTER 5
RESULTS OF FIELD STUDIES

Dose-response studies were performed during the first week of testing at each of the test sites. The purposes of the dose-response studies were to provide basic information on the effects of MIEX® dose and contact time, and to provide guidance on the range of doses and contact times to be evaluated during the four weeks of pilot study that followed. This testing was accompanied by characterizations of NOM properties in raw water and following different treatment simulations based on molecular weight, XAD fractionation, and THM/HAA formation. Coagulation testing was also conducted during this period at utilities where flocculation, clarification, and filtration are employed in the treatment sequence. A second round of coagulation and NOM characterizations was performed during the last week of the pilot test period to provide added information.

Detailed pilot testing was performed for the four weeks after the initial week of pilot plant start-up and bench-scale testing. Pilot plant operating results included evaluations of UV254 absorbance reduction and DOC removal by the MIEX® process, the effects of changes in contact time on MIEX® process performance, the effect of the MIEX® process on basic water quality conditions, and the formation of THMs and HAA. Special studies were also conducted at selected sites to evaluate the removal of sulfide, bromide, arsenic, and perchlorate.

Maytum Water Treatment Plant

The Maytum Water Treatment Plant (MWTP) is a 6 mgd facility owned by the City of New Port Richey, FL and operated in a cooperative arrangement with Tampa Bay Water. The principal water source for the plant is the Floridian Aquifer. This source has modest levels of DOC and can form levels of THMs and HAAs that are significant as compared with the MCLs. The treatment process includes aeration and chlorination. The testing at the MWTP was focused on DOC and sulfide removal, with particular attention to the effect of DOC removal on the formation of total THMs (TTHMs) and HAA5. Table 5.1 lists the raw water quality encountered during the testing, which took place from November 17 through December 21, 2002.

Table 5.1
Maytum WTP raw water quality

	Average	Minimum	Maximum
Temperature (°C)	24.8	23.0	26.0
pH	7.62	7.51	7.73
Turbidity (ntu)	0.10		
Total alkalinity (mg/L)	200	150	235
Total hardness (mg/L)	239	201	301
Conductivity (μmhos/cm)	451	414	475
Total sulfide (mg/L)	1.7	0.9	2.1
Chloride (mg/L)	10	5	15
DOC (mg/L)	1.93	1.58	2.54
TDS (mg/L)	247	223	265
Sulfate (mg/L)	17.9	6.5	24.9

In terms of constituents that are generally indicative of potential for competition for ion exchange sites, the alkalinity of the raw water was relatively high. Moderate levels of TDS were also encountered, but the sulfate and chloride levels were relatively low. The DOC concentration is relatively low, but remains a concern at this plant, because the plant's treatment train does not include provisions for DOC removal. The addition of chlorine within the current treatment process sequence can result in the formation of undesirable levels of THMs and HAAs as a consequence.

Bench-scale Testing

MIEX® dose-response testing was performed using the procedures described in Chapter 4. These procedures involve the addition of a blend of un-regenerated and freshly regenerated resin that are collected from the pilot plant after several days of operation to allow a steady state condition to be reached in these two resin streams. Resin from these two sources was blended to simulate a 5 percent regeneration ratio (95 percent un-regenerated resin from the recycle stream at the bottom of the sedimentation tank and 5 percent freshly regenerated resin). Sulfide, UV254, and DOC results from this testing are shown on Figures 5.1 through Figure 5.3.

The results of the dose-response testing indicated relatively high levels of DOC and UV254 reduction for a broad range of doses and detention times. The removals showed a tendency of decreasing benefit as doses and contact times were increased, with little change at higher MIEX® doses. This relationship suggests that some components of NOM were not susceptible to further removal by the MIEX® process. Sulfide removal was observed at all of the conditions tested. A slight increase in removal was observed when both dose and contact times were increased, but the removal percentages tended to be high at all conditions.

There was little difference in reductions of UV254 absorbing material, DOC, and sulfide between contact times of 15 and 20 minutes, whereas some difference was observed in UV254 and DOC removals at a contact time of 7.5 minutes and at lower resin doses. Effects of contact time were much less pronounced at resin doses higher than 15 mL/L.

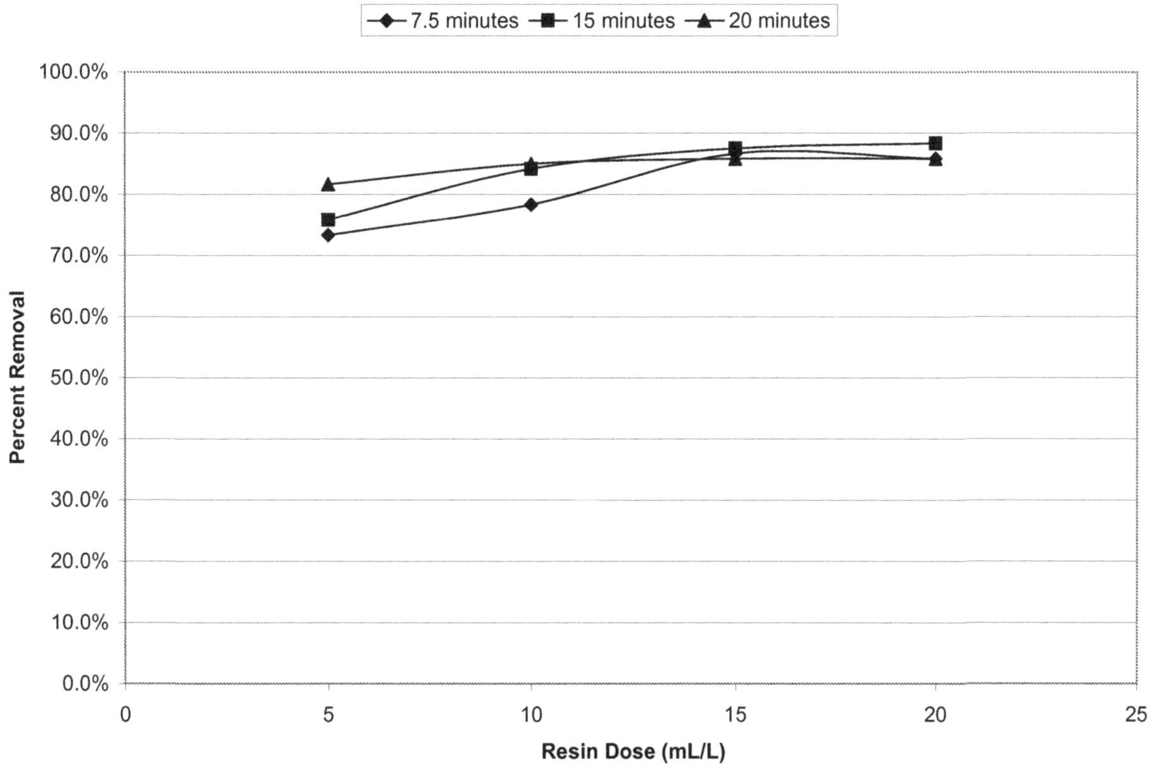

Figure 5.1 Percent UV254 absorbance reduction during dose-response tests at MWTP

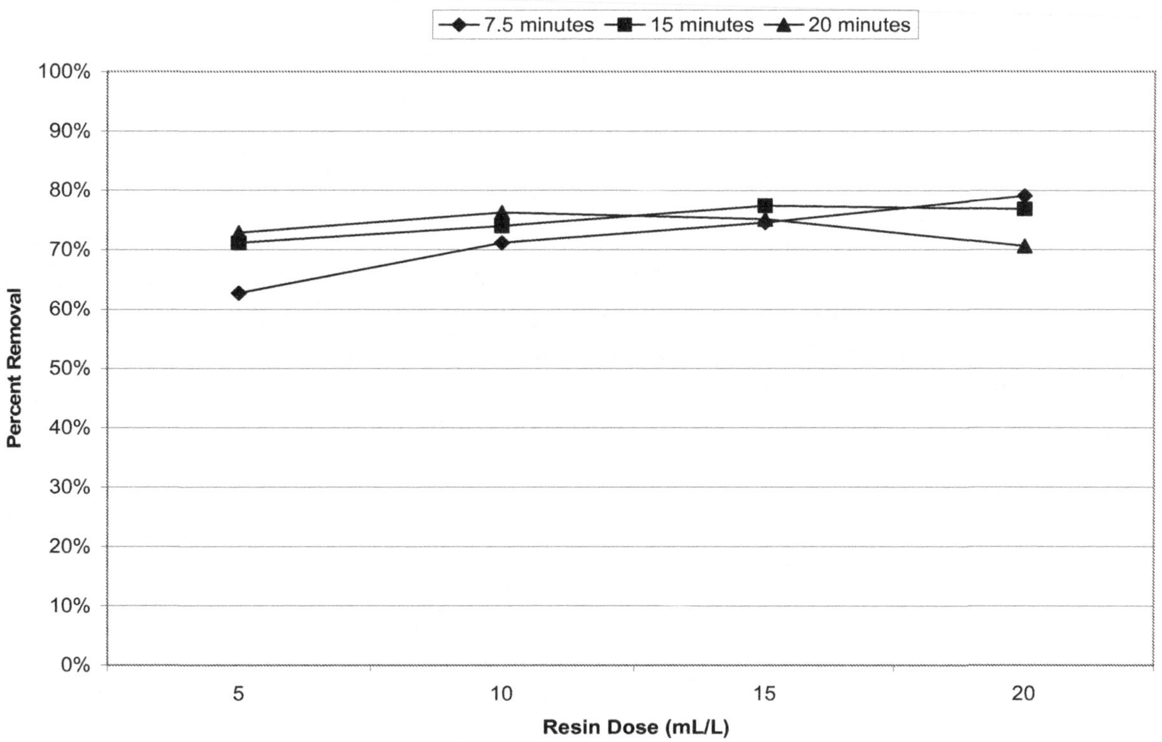

Figure 5.2 Percent DOC removal during dose-response tests at MWTP

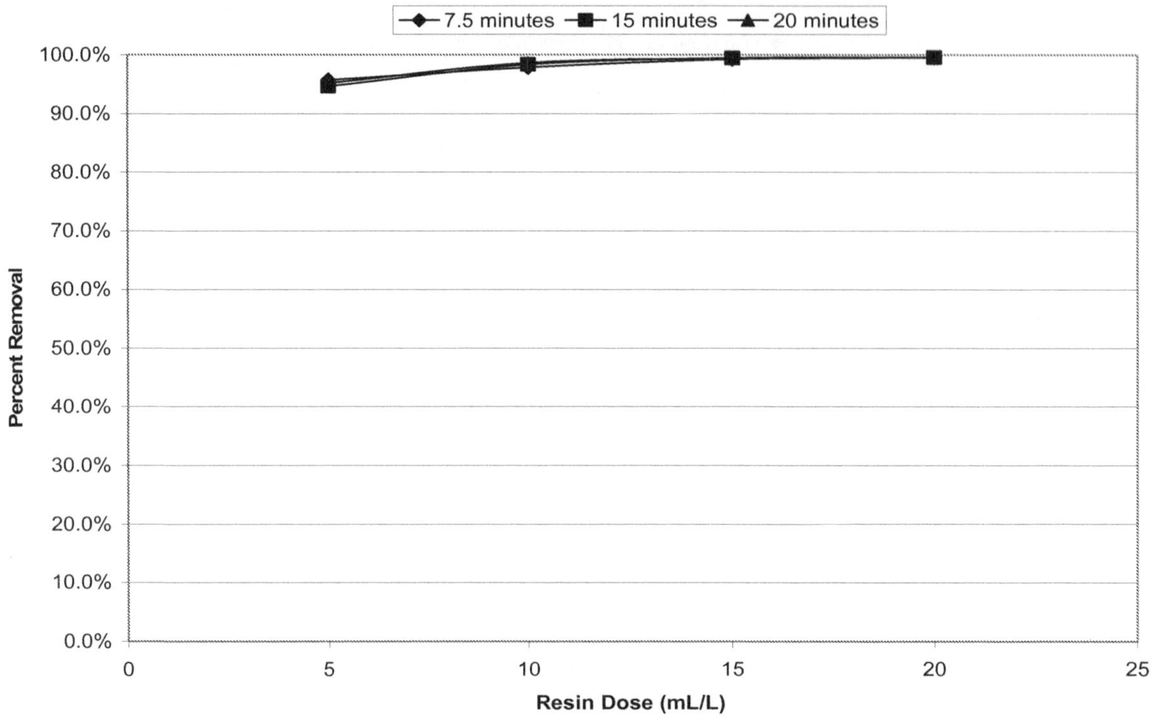

Figure 5.3 Percent total sulfide removal during dose-response tests at MWTP

Pilot Testing

Pilot testing was performed using resin concentrations of 12 and 20 mL/L, based on the results of the dose response testing that indicated slight deterioration in performance at doses below this range. There is also a need to maintain adequate resin in suspension to achieve efficient contacting in the settling tank, and these concentrations are in the range that proved workable at full scale at the Wanneroo Groundwater Treatment Plant; concentrations that are too low (less than 8 mL/L) can result in less efficient resin capture. The regeneration percentages used were 5 percent or less. These relatively low percentages were selected because the performance of the resin appeared to be reaching a point of diminishing return in both the bench scale dose response results and during the early stages of pilot testing as well. Sulfide, UV254, and DOC results from this testing are shown on Figures 5.4 through 5.6. (There is a gap in the testing period that corresponds to a break in operation during November when the plant was not being operated.)

Pilot operating conditions for each week of the four week test period are summarized below:

- Week 1: 20 mL/L resin concentration, 15 minute contact time, and 5 percent of flow sent to regeneration.
- Week 2: 12 mL/L resin concentration, 15 minute contact time, and 5 percent of flow sent to regeneration.
- Week 3: 12 mL/L resin concentration, 15 minute contact time, and 2.5 percent of flow sent to regeneration.
- Week 4: 20 mL/L resin concentration, 15 minute contact time, and 5 percent of flow sent to regeneration.

Resin concentration was changed by removing and storing resin when a reduction was required and adding resin from storage when an increase in concentration was required. Fresh resin was periodically added to the system to make up for losses.

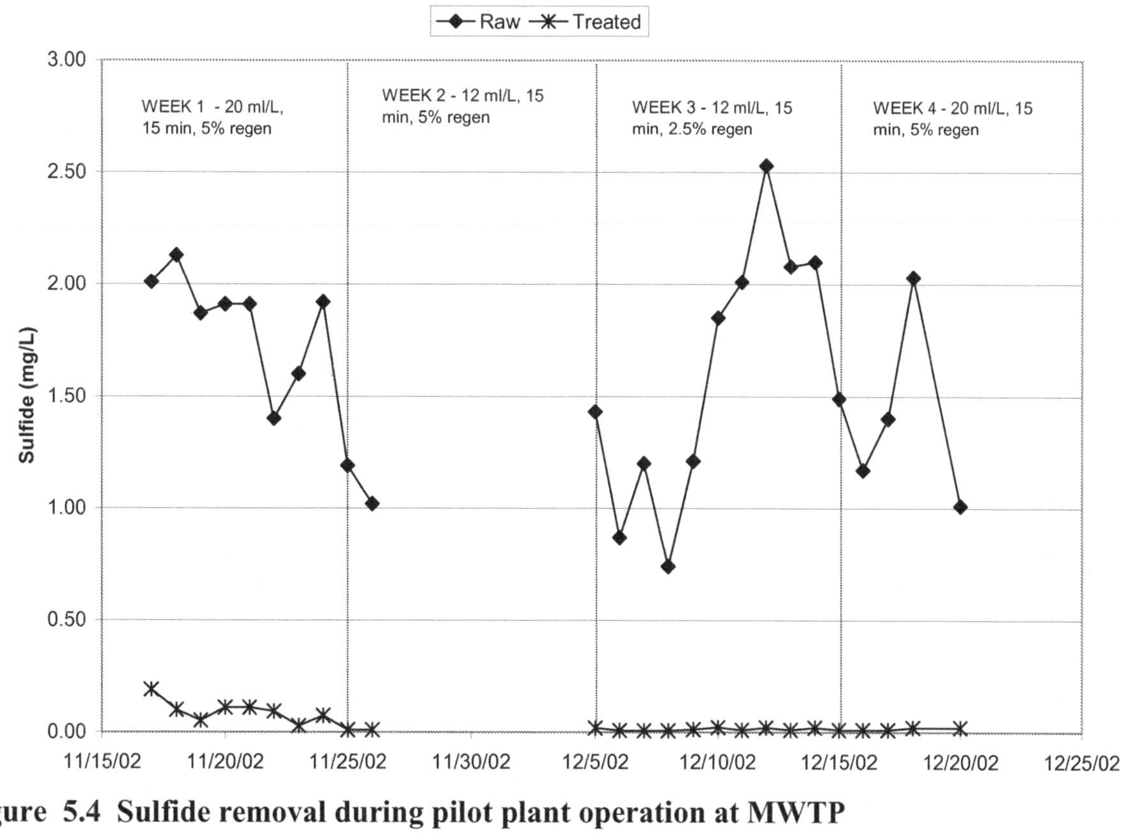

Figure 5.4 Sulfide removal during pilot plant operation at MWTP

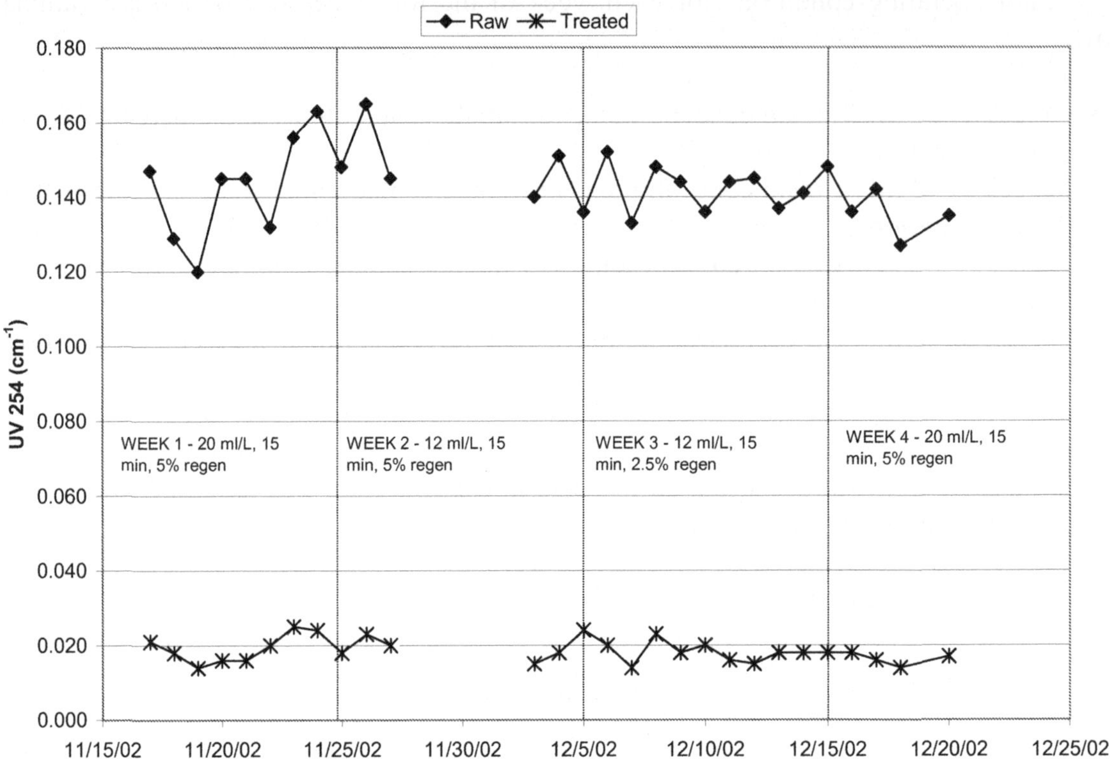

Figure 5.5 UV254 absorbance reduction during pilot plant operation at the MWTP

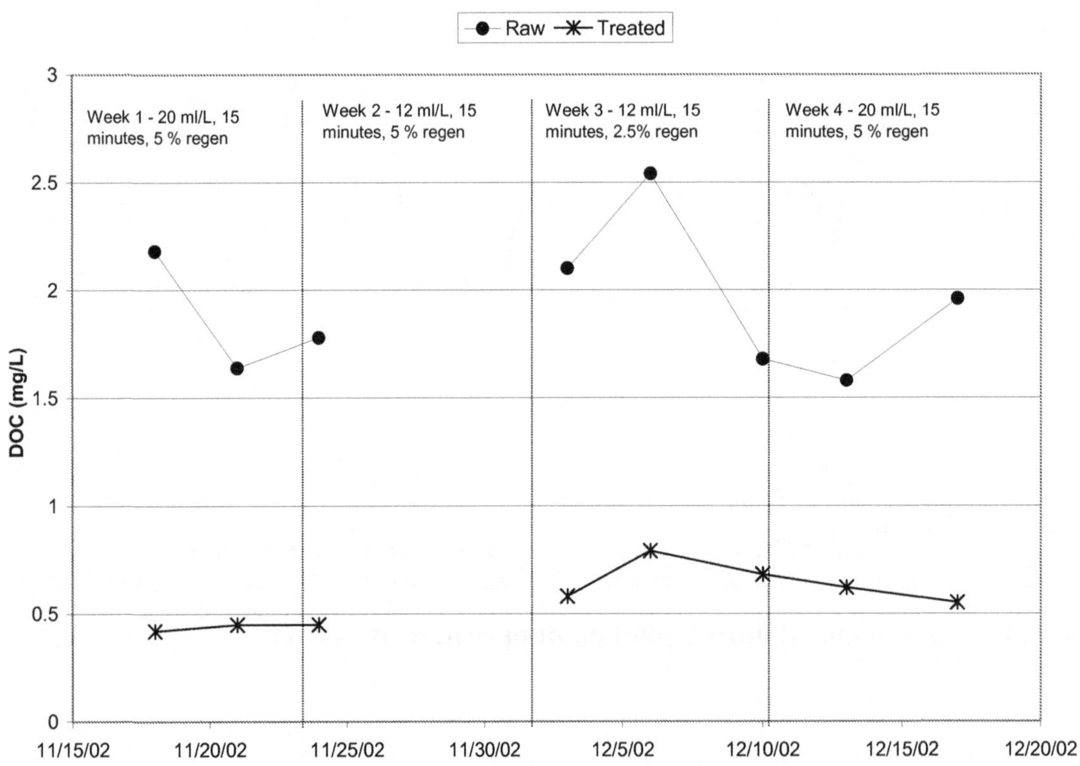

Figure 5.6 DOC removal during pilot plant operation at MWTP

These results indicated that the MIEX® process removed significant percentages of UV254, DOC, and total sulfide (Table 5.2). DOC removal was greater than 60 percent, and reduction in UV254 absorbance was more than 80 percent. Total sulfide removal was between 94 and 99 percent at all operating conditions and improved as the pilot operation progressed, regardless of feed water concentration. It is possible that the removal of sulfide by the MIEX® resin was augmented by biological activity. Colonies of filamentous bacteria were observed in the settling basin throughout the four weeks, and the colonies may have formed at the surface of the resin. In addition, the sulfide concentration in the regenerant brine was approximately 1 mg/L, a concentration that is close to the level found in the raw water. This is a very low concentration when the potential concentrating effect of the resin is taken into account.

Although UV254 absorbance reduction remained generally consistent throughout the pilot study, the DOC removal did decrease over time, which may have reflected a shift in the concentration and character of the DOC in the raw water rather than a decline in the effectiveness of the resin in terms of removing DOC. There were periods of extensive rainfall that resulted in visual changes in water characteristics during the test period and different combinations of wells, each with a slightly different water quality, are applied at this site. Therefore, variation in water quality is to be expected. Benchmark testing that compared the DOC removal results of a resin sample obtained at the end of the testing period with a resin sample from the beginning of the testing period showed little change in the resin's capacity to remove DOC. Therefore, progressive loss of resin capacity for DOC uptake is not the likely cause of observed changes in percent DOC removal.

Table 5.2
Percent removals of total sulfide, UV254 absorbance, and DOC at MWTP

	Week 1	Week 2	Week 3	Week 4
Total sulfide	94%	98%	99%	99%
UV254	87%	87%	86%	88%
DOC	77%	73%	65%	67%

Detention Time Study

The fact that different operating conditions did not have a significant effect on the removal of the different constituents at 15-minutes' detention time is consistent with the results of the bench-scale studies at detention times of 15 minutes and longer. Detention time studies were also performed to determine the effect of reducing the contact time to 7.5 minutes at the respective dosage and percent regeneration conditions used during each week of operation. This change was affected by bypassing one of the two contactors to reduce the contacting volume by half. A period of two hours was allowed for the change to take place across the pilot plant. The results of these studies, listed in Table 5.3, provide an indication of how contact time impacts process efficiency.

Table 5.3
Detention time study results at MWTP

	Week 1 - 20 mL/L, 5% regeneration		Week 2 - 12 mL/L , 10% regeneration		Week 3 - 12 mL/L, 2.5% regeneration		Week 4 - 20 mL/L, 5% regeneration	
	Percent removal 7.5 minutes	Percent removal 15 minutes	Percent removal 7.5 minutes	Percent removal 15 minutes	Percent removal 7.5 minutes	Percent removal 15 minutes	Percent removal 7.5 minutes	Percent removal 15 minutes
UV254 cm-1	85	87	73	87	78	86	77	88
DOC mg/L	71	73	60	72	54	60	61	72
Sulfide mg/L	86	94	82	98	67	99	93	99

Evaluations of THM and HAA5 Formation

The effects of NOM removal by the MIEX® process on the formation of THMs and HAA5 were evaluated during two periods: at the beginning of testing and during the last week of pilot testing. Disinfection by-product formation took place under uniform formation conditions (UFC) with all samples incubated at a temperature of 20 °C and a pH of 8.0. Raw water for the testing was obtained at the pilot plant influent, and the pilot plant effluent was used as the treated water supply. The MIEX® process was operated at a contact time of 15 minutes, a regeneration rate of 5 percent, and a resin concentration of 20 mL/L during both testing periods. Raw and MIEX® treated DOC levels were 1.6 and 0.5 mg/L, respectively, for the initial test period, and 2.0 and 0.6 mg/L, respectively, for the final test period. The resulting THM and HAA5 formation values are shown in Tables 5.4 through 5.7.

These results indicate that using the MIEX® process to treat raw water results in significant reductions in THM and HAA formation. Chloroform was the dominant THM formed in both sets of tests, and the primary effect of reducing NOM with the MIEX® process on THMs was a reduction in the chloroform concentration. However, no general reduction was observed in the ultimate formation of the brominated THM species; instead, there was a general increase in the 3-day formation levels of these species. Therefore, while the MIEX® process achieved a gross reduction of THMs in this case, it was not effective for removing the brominated species. This may not be a consideration for this test site, where the level of formation of brominated species is relatively low; however, this may become more of a consideration at other locations. Note this shift in speciation to more brominated forms is also observed with other processes that remove DOC and is not confined to MIEX alone. The effects on speciation of brominated HAAs were not identified at this location, as formation of these DBPs was below detection limits in both raw and MIEX® treated samples.

Table 5.4
THM concentrations from DBP formation potential experiments at MWTP – Week 1

	Chlorine residual (mg/L)			Cl_3CH (µg/L)			$BrCl_2CH$ (µg/L)			Br_2ClCH (µg/L)			Br_3CH (µg/L)			Total THM (µg/L)		
	2 hour	1 day	3 day	2 hour	1 day	3 day	2 hour	1 day	3 day	2 hour	1 day	3 day	2 hour	1 day	3 day	2 hour	1 day	3 day
Raw water	2.1	1.0	0.2	37.5	97.9	98.1	6.1	13.1	11.8	<2	<2	<2	<2	<2	<2	43.6	111.0	109.9
Post MIEX®	1.6	1.1	0.5	8.7	34.0	45.3	2.7	9.8	13.7	1.4	3.9	5.3	<2	<2	<2	12.8	47.7	64.3

Table 5.5
THM concentrations from DBP formation potential experiments at MWTP – Week 4

	Chlorine residual (mg/L)			Cl_3CH (µg/L)			$BrCl_2CH$ (µg/L)			Br_2ClCH (µg/L)			Br_3CH (µg/L)			Total THM (µg/L)		
	2 hour	1 day	3 day	2 hour	1 day	3 day	2 hour	1 day	3 day	2 hour	1 day	3 day	2 hour	1 day	3 day	2 hour	1 day	3 day
Raw Water	2.2	1.0	0.3	38.3	99.7	138.0	5.6	12.7	16.1	0.6	0.9	1.0	<2	<2	<2	44.5	113.3	155.1
Post MIEX®	1.6	1.4	1.1	3.5	12.9	22.8	2.1	7.5	11.8	1.3	3.8	5.5	<2	<2	<2	6.9	24.2	40.1

Table 5.6
HAA5 concentrations from DBP formation potential experiments at MWTP- Week 1

	ClAA (µg/L)			BrAA (µg/L)			Cl_2AA (µg/L)			Br_2AA (µg/L)			Cl_3AA (µg/L)			Total HAA5 (µg/L)		
	2 hour	1 day	3 day	2 hour	1 day	3 day	2 hour	1 day	3 day	2 hour	1 day	3 day	2 hour	1 day	3 day	2 hour	1 day	3 day
Raw water	2.3	3.7	<2	<2	<2	<2	10.5	18.3	25.9	<2	<2	<2	19.4	27.3	30.6	32.2	49.3	56.5
Post MIEX®	<2	<2	<2	<2	<2	<2	<2	2.6	4.2	<2	<2	<2	3.5	3.5	4.3	3.5	6.1	8.5

Table 5.7
HAA5 concentration from DBP formation potential experiments at MWTP – Week 4

	ClAA (µg/L)			BrAA (µg/L)			Cl_2AA (µg/L)			Br_2AA (µg/L)			Cl_3AA (µg/L)			Total HAA5 (µg/L)		
	2 hour	1 day	3 day	2 hour	1 day	3 day	2 hour	1 day	3 day	2 hour	1 day	3 day	2 hour	1 day	3 day	2 hour	1 day	3 day
Raw water	<2	3.3	<2	<2	<2	<2	9.6	17.0	25.7	<2	<2	<2	15.6	25.1	32.1	25.2	45.4	57.8
Post MIEX®	<2	4.3	2.4	<2	<2	<2	<2	4.4	5.7	<2	<2	<2	3.5	6.5	6.4	3.5	15.2	14.5

Analyses of Resin Activity and Buildup of DOC in the Regenerant Brine

Testing was performed to compare the resin activity in samples of virgin resin, regenerated resin collected when the pilot plant was started up, and regenerated resin collected at the end of the pilot testing period after five weeks of operation. The purpose of these tests was to determine whether pilot testing results over the course of the test period could have changed as a consequence of changes in resin activity. These evaluations were performed by conducting parallel dose-response tests to assess uptake of UV254 absorbance and DOC from a common raw water sample that was obtained from the test site during pilot testing. This testing differed from the previous dose response testing in that all of the resin tested for relative activity was in a freshly regenerated state; a mix of fresh and recycled resin was used in the previous tests. Therefore, these results are not indicative of process potential, but are structured to provide a relative measure of resin activity for removing DOC and reducing UV 254 from a common raw water source.

The concentration of DOC in the test water was 2.1 mg/L and the UV254 absorbance was 0.116 cm^{-1}. Results from these comparative dose-response tests are summarized in Table 5.8. These results show good agreement in the activity of the respective resin samples in terms of uptake of both DOC and UV254 absorbance at each of the MIEX® doses that are applied and indicate that there was very little change in the activity of the resin for NOM removal over the course of the testing.

Table 5.8
Evaluations of resin activity at MWTP

Utility	MIEX® dose (mL/L)	UV254 absorbance reduction (%)	DOC removal (%)
Virgin resin	1	64	42
	2	81	59
	4	93	76
Regenerated resin at the beginning of operation (one week of startup)	1	55	49
	2	68	65
	4	85	79
Regenerated resin at the end of five weeks of operation (four weeks of operation)	1	55	49
	2	73	62
	4	88	78

The condition of the regenerant solution is another consideration; the regenerant was re-used throughout each week of pilot testing. Brine was prepared as a salt concentration of 125 g/L and was recovered at the end of each regeneration cycle. Salt was added to the regenerant brine solution to make up for any loss of chloride during successive regenerations. One of the key considerations in the application of regenerant solution on a reuse basis is the extent of buildup of NOM. This was assessed by determining the DOC concentration in the brine at the end of each week of operation. Results of these analyses are shown in Table 5.9. These DOC increases are relatively low as compared with those at full scale at the WGTP in Western Australia, which can be 8 g/L or higher and reflect a much lower level of DOC in the raw water at this location.

Table 5.9
DOC buildup in regenerant brine at MWTP

Week	DOC concentration in regenerant brine at end of week (g/L)
1	0.6
2	1.4
3	0.9
4	1.3

Hanahan Water Treatment Plant

The HWTP is located in Hanahan, SC, and is owned and operated by the Charleston Commissioners of Public Works (CPW) and supplies water to the city of Charleston. The plant receives water from two sources: the Edisto River and the Bushy Park Reservoir. Raw water from the Edisto River has an average TOC concentration of 10 mg/L (range 2.4 to 19.1 mg/L) and an average alkalinity of 22 mg/L as $CaCO_3$. Bushy Park Reservoir water has an average TOC of 6 mg/L (range 3 to 13 mg/L) and an average alkalinity of 28 mg/L as $CaCO_3$. Treatment consists of rapid mixing, flocculation, sedimentation, and granular media filtration. Alum is used as the primary coagulant and free chlorine as the primary disinfectant. Chloramines are used to provide a finished water disinfectant residual in the distribution system. Testing of the MIEX® process at the HWTP focused on removal of DOC and bromide, lowering coagulant dosages, and reducing disinfection by-products. The testing was conducted from January 25 through March 1, 2003.

Table 5.10
Hanahan WTP raw water quality

	Average	Minimum	Maximum
Temperature (°C)	7.8	6.0	10.0
pH	7.25	7.01	7.44
Turbidity (ntu)	3.16	2.79	4.20
UV254 (cm^{-1})	0.217	0.203	0.236
DOC (mg/L)	5.01	4.60	5.33
SUVA	4.3	--	--
Total alkalinity (mg/L)	27	25	29
Total hardness (mg/L)	31	23	64
Bromide (mg/L)	0.09	0.05	0.12
Sulfate (mg/L)	14.5	9.6	15.5
Chloride (mg/L)	17.5	9.2	20.0
Conductivity (μmhos/cm)	162	160	170
TDS (mg/L)	125	84	260

Raw water characterisctics during the period of testing are summarized in Table 5.10. Because the raw water contained relatively low concentrations of TDS and correspondingly low concentrations of chloride and sulfate, as well as alkalinity, the inorganic constituents that could compete with DOC and other anions for uptake on the resin were limited. DOC concentrations were moderately high, and bromide was present at significant concentrations in terms of ability to affect the formation of disinfection by-products. Water temperatures during the testing ranged from 6 to 10°C.

Bench-scale Testing

The results of bench-scale MIEX® dose-response testing for reduction of UV254 absorbance and removal of DOC are shown on Figures 5.7 and 5.8. These results indicate a greater impact of both detention time and resin dose on the reduction/removal of these constituents than observed at MWTP. Similar to the bench scale testing results at the MWTP, the fact that the maximum reduction in UV254 absorbance and DOC tended to approach a limit with increased dose and contact time gave an indication that the water contained some organic material that could not be readily removed by the resin. As shown in Table 5.11, the results of bromide analyses conducted during MIEX® dose-response testing at this site suggest that bromide can be removed if adequate detention time and dose are provided.

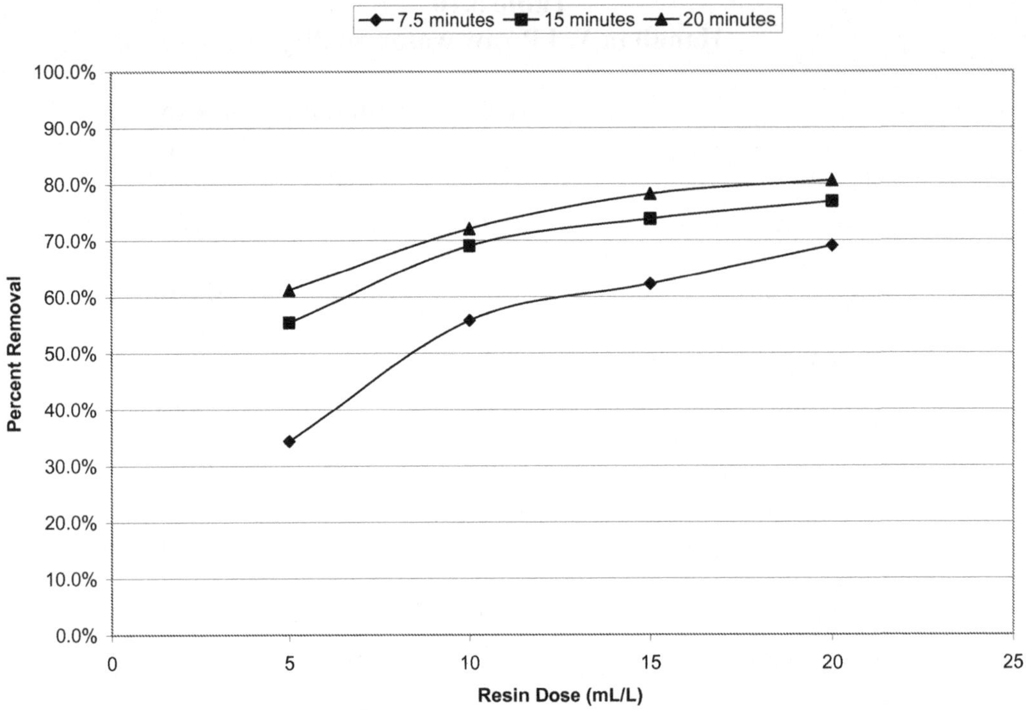

Figure 5.7 Percent UV254 absorbance reduction during dose-response tests at HWTP

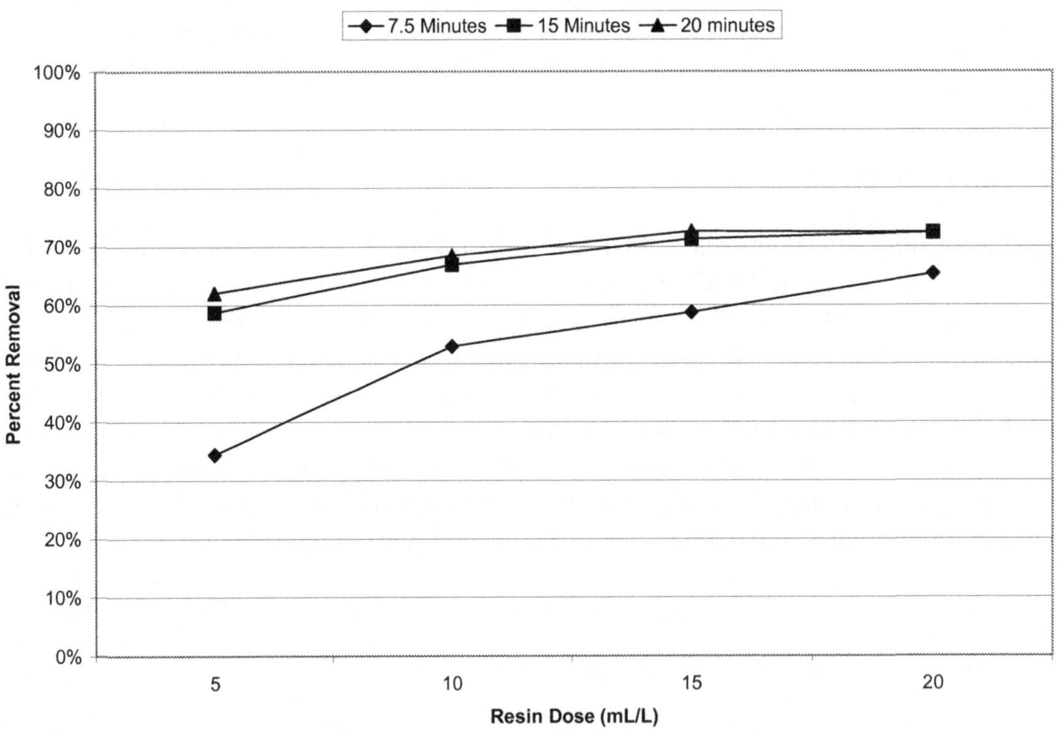

Figure 5.8 Percent DOC removal during dose-response tests at HWTP

Table 5.11
Bromide analysis results from dose-response tests at HWTP (mg/L)

95 percent preloaded resin and 5 percent fresh resin	7.5 minutes	15 minutes	20 minutes
Raw water	0.09	0.09	0.09
5 mL/L	0.07	< 0.05	< 0.05
10 mL/L	< 0.05	< 0.05	< 0.05
15 mL/L	< 0.05	< 0.05	< 0.05
20 mL/L	0.07	< 0.05	< 0.05

Pilot Testing

Pilot testing was performed with a resin concentration of 20 mL/L, regeneration rates of up to 10 percent, and contact times of up to 20 minutes. Specific operating conditions during the four consecutive weeks of pilot testing at the Hanahan WTP are summarized below:

- Week 1 - 20 mL/L resin, 20 minute contact time, and 5 percent of flow sent to regeneration.
- Week 2 - 20 mL/L resin, 20 minute contact time, and 10 percent of flow sent to regeneration.
- Week 3 - 20 mL/L resin, 15 minute contact time, and 5 percent of flow sent to regeneration.
- Week 4 - 20 mL/L resin, 20 minute contact time, and 5 percent of flow sent to regeneration.

During Week 4, a decision was made to use a rock salt source for preparing the brine, in place of the solar salt that has been used in the past. This step was taken to determine whether the salt composition was affecting the extent of bromide removal. (Solar salt is generally obtained from evaporation of sea water, and may contain bromide at levels that could impede the ability to restore resin capacity for bromide uptake during regeneration.)

UV254 absorbance reductions during the four weeks of pilot operation are shown on Figure 5.9. These reductions were affected by changes in contact time and percent regeneration. The highest removal percentage occurred during Week 2, when the longest contact time and the highest regeneration percentage were used. These results differ from those obtained during the testing at the MWTP, where varying operating conditions did not produce a discernable effect except at the shorter detention time of 7.5 minutes. This difference between locations may be attributable to the higher concentrations of NOM in the raw water supply or slower kinetics at the Hanahan WTP, and suggests that the criteria for contact time will vary depending on site-specific conditions. The highest DOC removals were also attained during Week 2 of testing as shown on Figures 5.10 and 5.11. There was a degree of reduction in DOC removal during Week 4 as compared with results obtained for similar conditions of MIEX process operation during Week 1. Resin activity testing (see Table 5.20) indicated little change in DOC uptake capability over the course of testing at this site. Therefore, changes in resin properties are not a likely cause and it is more likely that the change occurred as a consequence of changes in raw water properties.

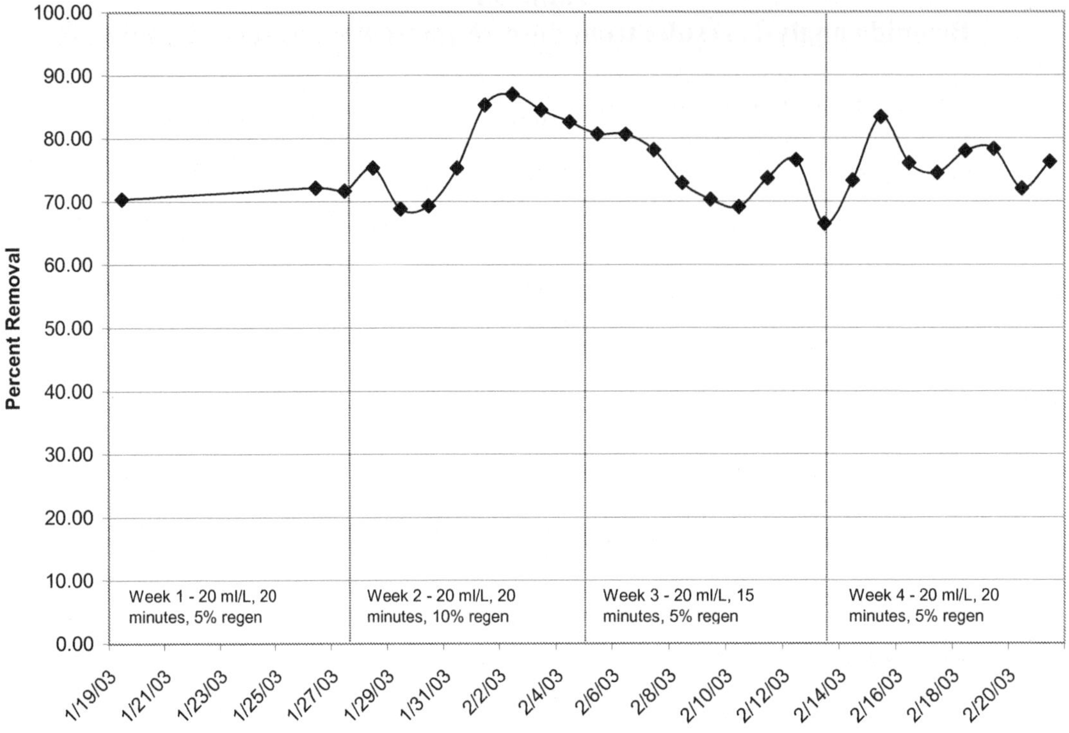

Figure 5.9 Percent reduction of UV254 absorbance during pilot plant operation at HWTP

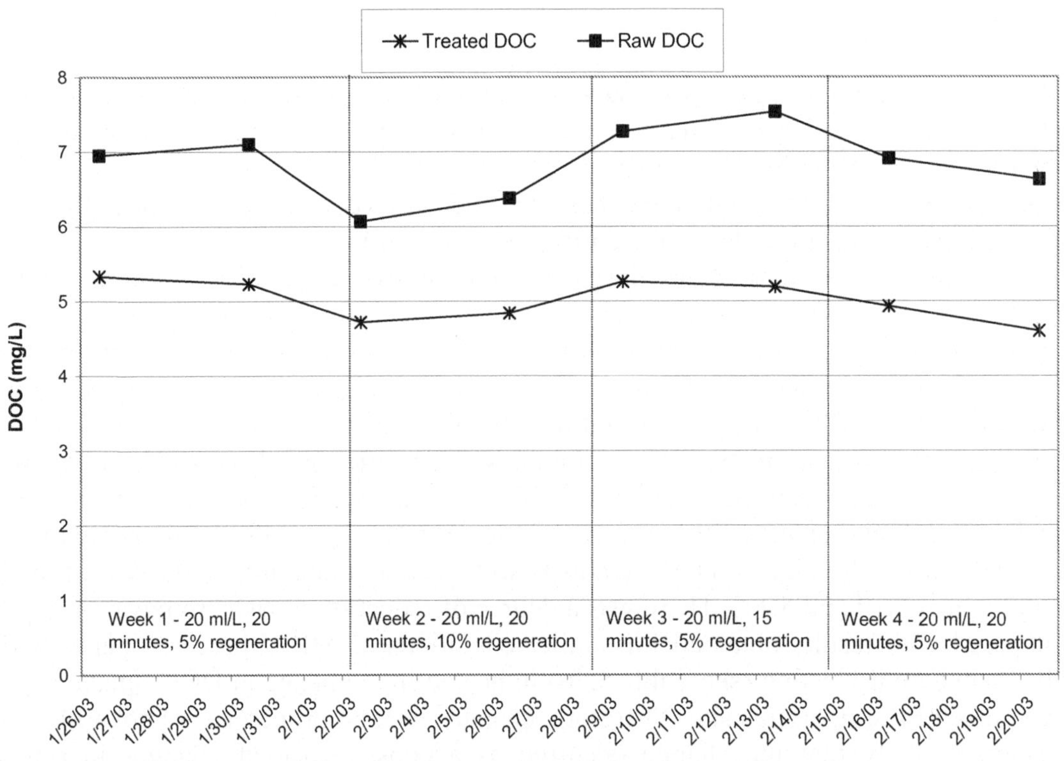

Figure 5.10. DOC results during pilot plant operation at HWTP

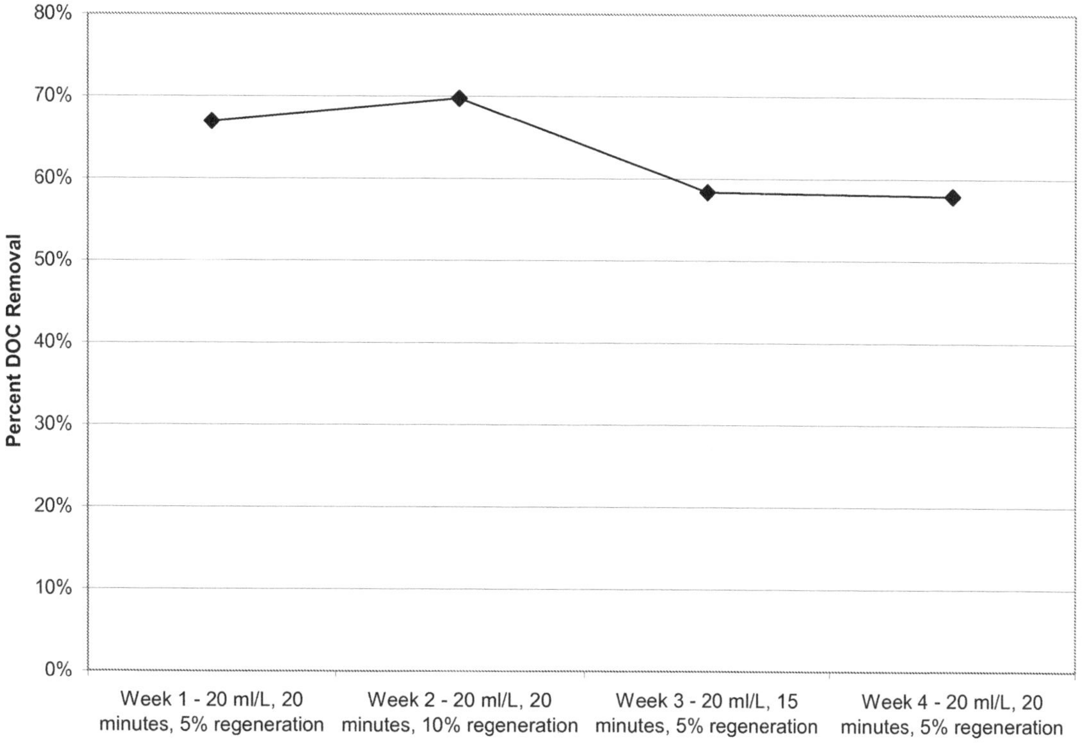

Figure 5.11 Percent DOC removals during pilot plant operation at HWTP

Bromide removal results are shown on Figure 5.12. These results are affected by the conditions of regeneration. No bromide removal was achieved during Week 1 at a regeneration rate of 5 percent when solar salt was used to prepare the brine. During Week 2, bromide removal was achieved only on several days when the pilot unit was operated at a regeneration ratio of 10 percent. Bromide removal was limited during Week 3 when the regeneration ratio was reduced to 5 percent, but was re-established when rock salt was substituted for the sea (solar) salt that had been used for preparing the regenerant. The consistent removal of bromide that was established with the use of rock salt during Week 4 raises the potential for improving bromide removal with the MIEX® process through the selection of salt source for makeup of regenerant brine. This consideration was not a goal of the original study plan and deserves further study to develop more definitive conclusions.

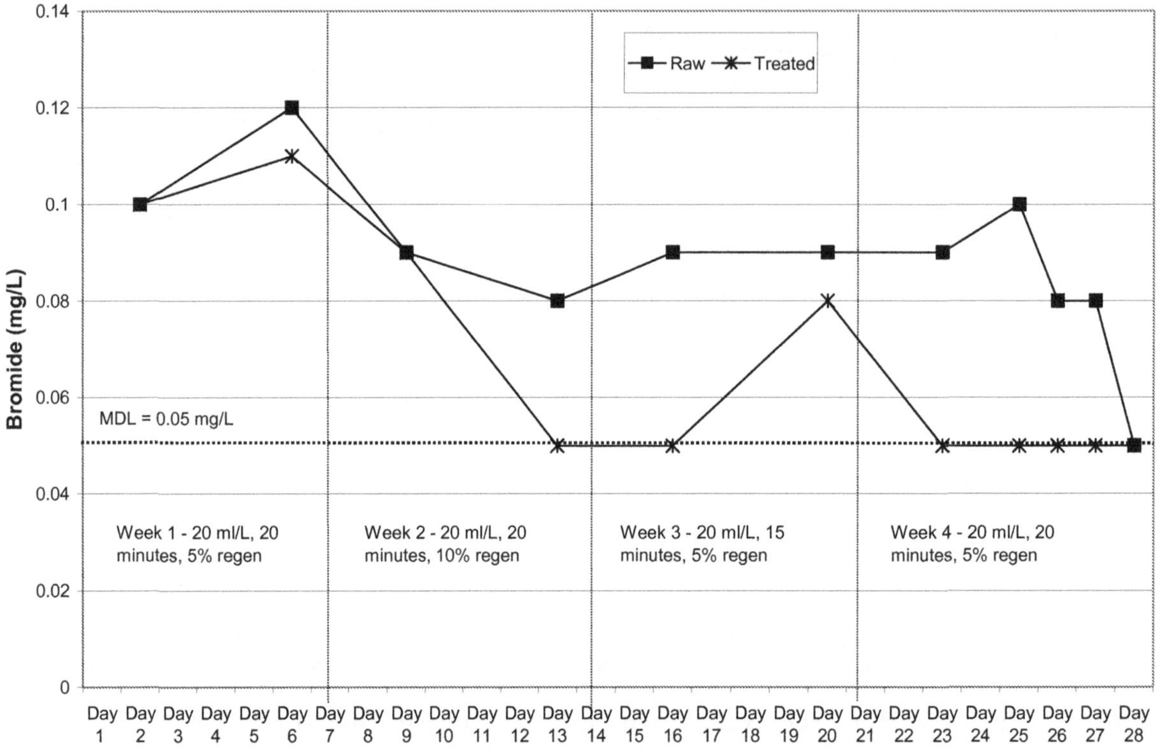

Figure 5.12. Bromide results during pilot plant operation at HWTP

As an additional consideration in pilot testing at HWTP, heterotrophic plate count analyses were performed on raw and MIEX® treated water during Week 4 of the pilot study (February 17, 2003). The results showed an increase in plate counts from 200 CFU in the raw water to 1,200 CFU in the treated water. These results indicate that some biological activity was occurring within the MIEX® process. However, this level of increase should not represent a problem as these types of bacteria are typically eliminated by modest levels of disinfection.

Detention Time Study

Detention time studies were performed each week to determine the effect of decreasing the contact time by half. This change was effected by bypassing one of the two contactors to reduce the contacting volume by half. A period of two hours was allowed for the change to take place across the pilot plant. The results of these studies and the percent removals achieved at the contact time selected for the respective week of operation are listed in Table 5.12. These results indicate that the potential effects of changes in contact time.

Table 5.12
Results of detention time studies at HWTP

	Week 1 - 20 mL/L 5% regeneration		Week 2 - 20 mL/L 10% regeneration		Week 3 - 20 mL/L 5% regeneration		Week 4 - 20 mL/L 5% regeneration	
	Percent removal 10 minutes	Percent removal 20 minutes	Percent removal 10 minutes	Percent removal 20 minutes	Percent removal 7.5 Minutes	Percent removal 15 minutes	Percent removal 10 minutes	Percent removal 20 minutes
UV254 cm^{-1}	54	72	71	83	55	72	59	77
DOC mg/L	56	64	61	68	50	55	46	56

Effect of MIEX® Process on General Water Quality Characteristics

The MIEX® process is expected to have some effect on several water quality characteristics due to the exchange of a broad range of anions in solution for chloride. Table 5.13 summarizes monitoring results for several key parameters conducted during the pilot testing at this location.

Table 5.13
Average concentrations of constituents at HWTP after MIEX® treatment

	pH	Total alkalinity (mg/L)	Total hardness (mg/L)	Sulfate (mg/L)	Chloride (mg/L)	TDS (mg/L)
Raw	7.25	27	31	14.5	17.5	125
Week 1	7.27	27	46	10	35	120
Week 2	7.11	25	31	4	121	130
Week 3	7.25	26	26	8	51	100
Week 4	7.43	26	25.5	7	45	120

A reduction in sulfate concentrations was observed, as expected. Sulfate is a significant anion in most natural waters and has a high affinity for anion exchange sites due to its divalent charge. Chloride concentrations increased, reflecting the exchange for anions that are being taken up by the resin. TDS generally did not change significantly, reflecting the tendency for anions that are removed by the resin to be balanced somewhat by a corresponding release of chloride. Hardness, alkalinity, and pH showed little change. The slightly higher hardness in the MIEX® treated water during week 1 of pilot operation reflects an increase in raw water hardness during this period. Figures 5.13 through 5.15 show the trends for sulfate, chloride, and TDS during the four weeks of pilot testing. The greatest reduction in sulfate concentrations and the greatest increase in chloride concentrations occurred during Week 2 when the highest regeneration ratio of 10 percent was used. The increase in chloride concentration across the process ranged from 20 to 25 mg/L, except during week 2, when a large concentration increase was observed in a single spike. Changes in TDS were also observed for this event. As this increase coincided with the time of elevated chloride, these extreme observations could have resulted from a carry through of some brine from regeneration.

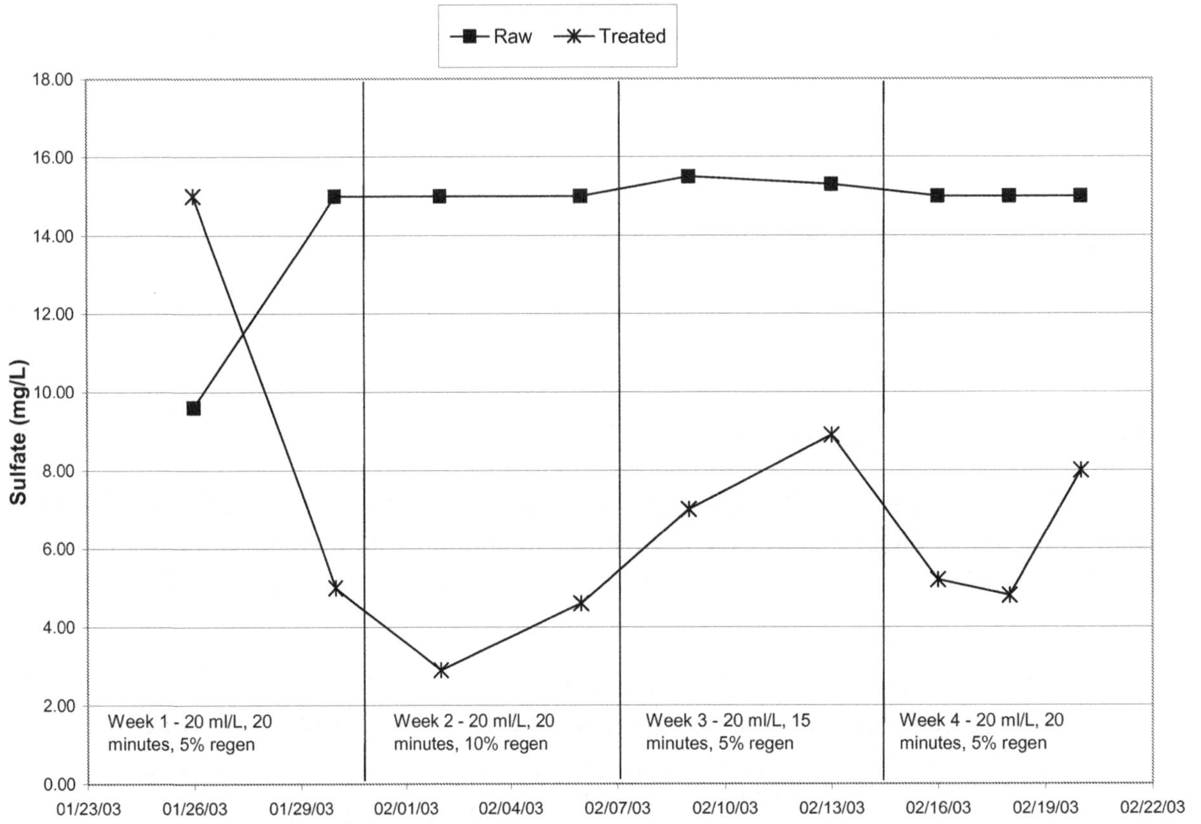

Figure 5.13 Sulfate concentrations during pilot plant operation at HWTP

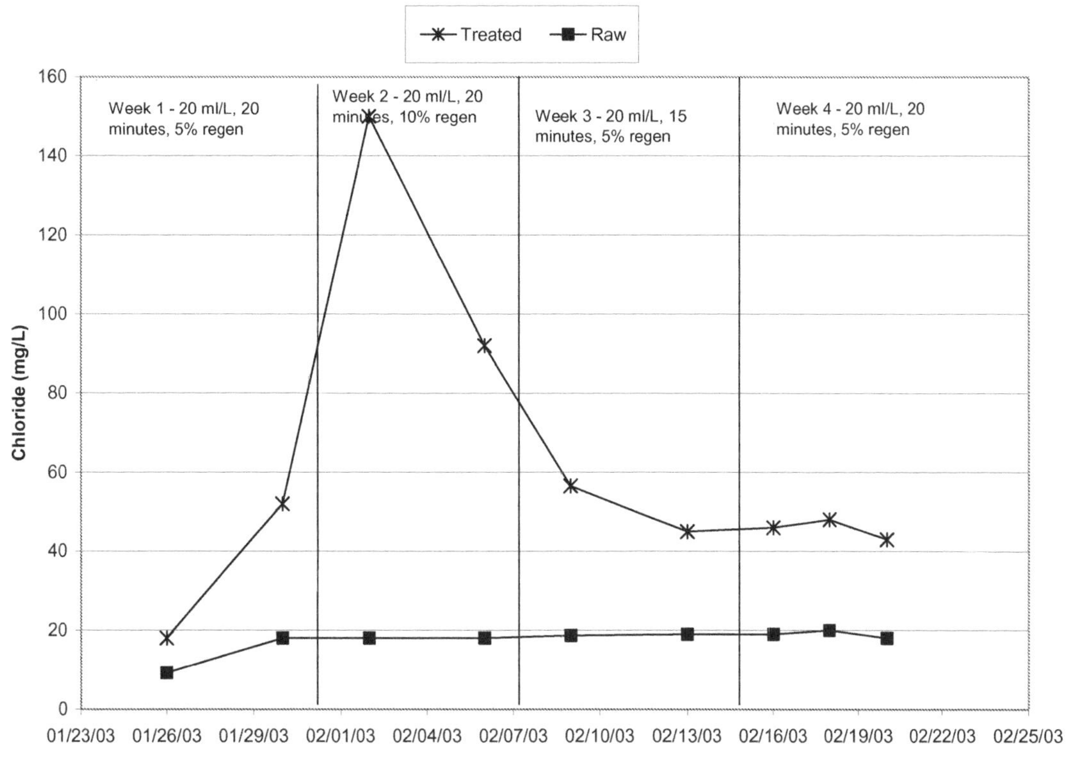

Figure 5.14 Chloride concentrations during pilot plant operation at HWTP

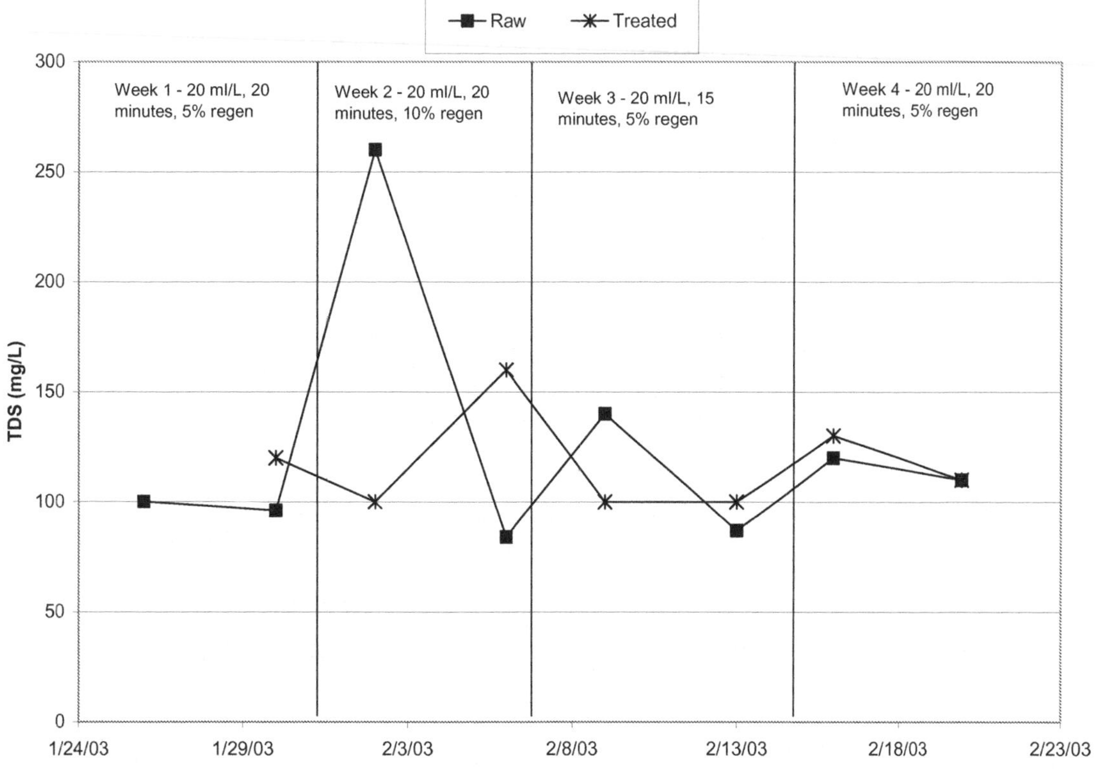

Figure 5.15 TDS concentrations during pilot plant operation at HWTP

Coagulation Study

A first set of bench scale coagulation studies was performed following completion of the dose-response testing initially conducted at the HWTP. The test sequence replicated the MIEX® process followed by coagulation in order to simulate operation of the full-scale plant retrofitted with the MIEX® process in this sequence. The MIEX® pilot plant was operated under conditions based on dose-response results, and the treated water from the pilot plant was collected for use in the bench scale coagulation testing. A constant temperature bath was used to match the temperature of the test water to the water temperature in the full-scale plant.

The same testing sequence was used for a second set of coagulant assessments at the end of the pilot study. Conditions for the first and second set of tests were similar, except that the brine used during the second set of tests was prepared with rock salt rather than with the solar salt that was used during the first set of tests. The use of the rock salt coincided with a significant reduction in finished water bromide concentrations as has been discussed.

The bench-scale tests were conducted using the MIEX® pilot plant effluent treated with 20 mL/L resin at a 20 minute contact time and a 5 percent regeneration rate. These tests simulated a MIEX®–coagulation sequence, and were conducted in parallel with a set of control tests using coagulated water without MIEX® pretreatment. Sodium hydroxide was added as required to adjust the pH to provide a spectrum of alum dose and pH conditions. Coagulation performance for the control and the MIEX®-coagulation tests was assessed from profiles in which the effect of both coagulant dose and pH on turbidity and UV254 absorbance was determined. Figures 5.16 and 5.17 show settled water turbidity and UV254 results for the first period of testing in a statistically fitted contour plotting format.

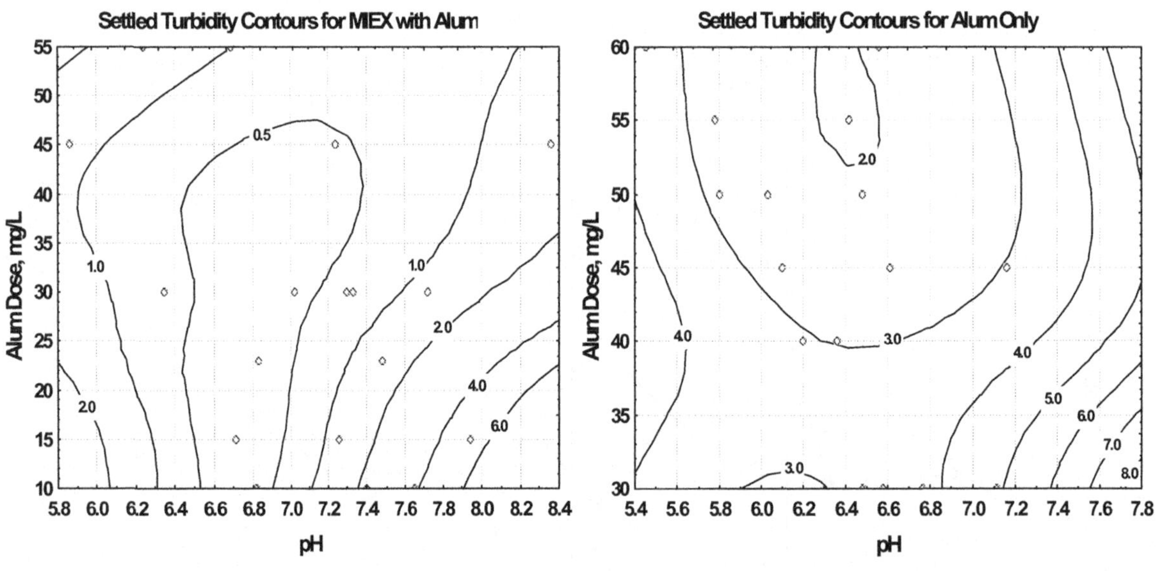

Figure 5.16 Settled water turbidity from coagulation evaluations during Week 1 at HWTP

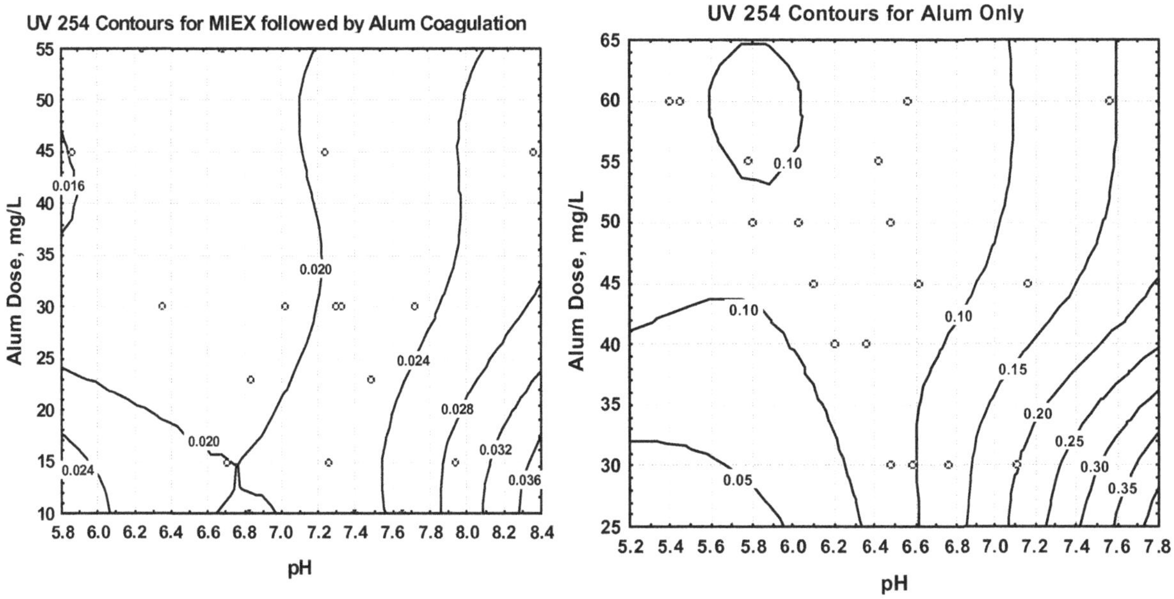

Figure 5.17 Settled water UV254 absorbance from coagulation evaluations during Week 1 at HWTP

These figures show significant differences in the effect of alum dose and pH on settled water turbidity and UV254 absorbance between MIEX®-pretreated and un-pretreated raw water. The coagulation process was enhanced by MIEX® pretreatment, as indicated by settled water turbidities of less than 1 ntu over a wide range of alum dosages and pH conditions, with alum doses of 10 mg/L or less capable of achieving excellent removal of turbidity. In contrast, the optimum coagulant condition for turbidity removal in the alum-only test was 55 mg/L at a pH of approximately 6.2 to 6.6. This is a very narrow optimum operating range that would require close operational supervision to maintain consistent process performance. Much lower settled water turbidities were possible with the MIEX®–alum treatment sequence than for any condition with alum only.

Comparisons of the UV254 results indicate substantial reductions when MIEX® resin is applied in conjunction with alum. These reductions are achieved over a much broader range of coagulant doses and pH than was the case for alum only.

Coagulation results for the second testing period were generally similar to those obtained during the first testing period, as shown on Figures 5.18 and 5.19.

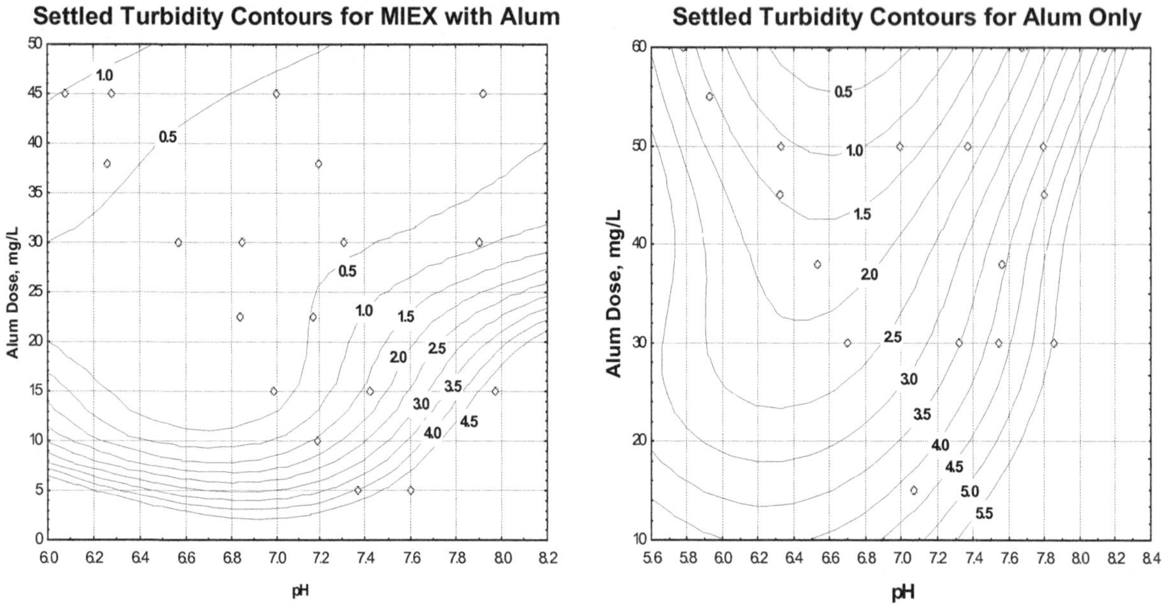

Figure 5.18 Settled water turbidity from coagulation evaluations during Week 4 at HWTP

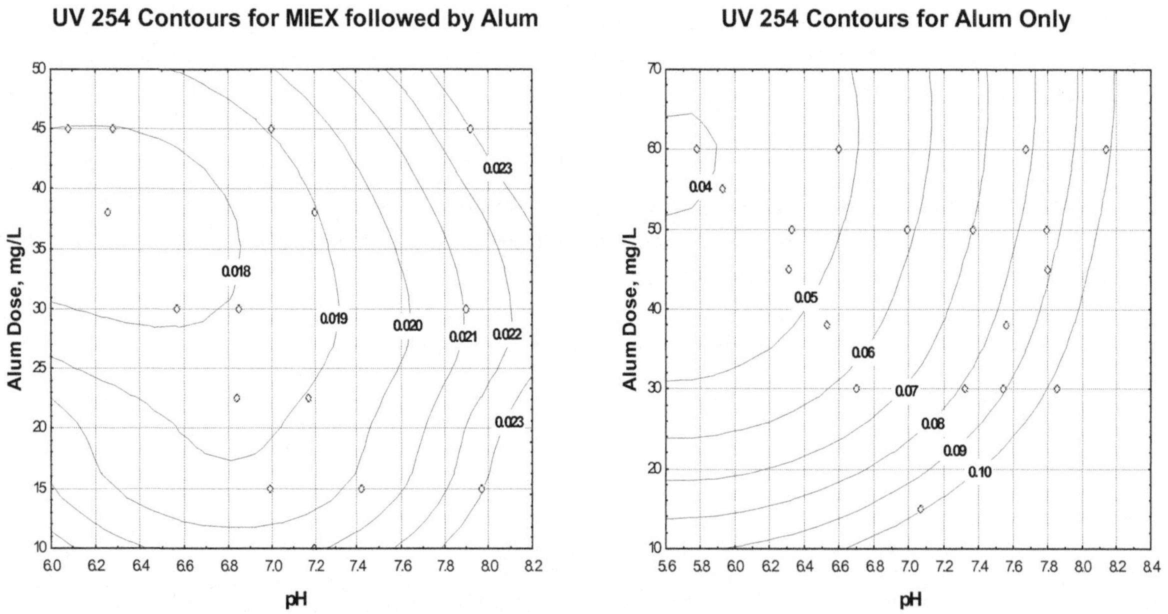

Figure 5.19 Settled water UV254 absorbance from coagulation evaluations during Week 4 at HWTP

Analyses of THM and HAA5 Formation

The effects of NOM removal by MIEX® resin and by the MIEX®–alum coagulation sequence on the formation of THMs and HAA5 were evaluated during two periods. Alum doses for evaluation of THM and HAA5 formation under the alum only and MIEX®–alum coagulation alternatives were selected based on the results shown in Figures 5.16 through 5.19. The alum only coagulation alternative was evaluated using an alum dose of 55 mg/L at a coagulation pH level of 6.0. Such conditions tend to maximize the effectiveness of alum for removal of NOM and are in a range that would be characterized as enhanced coagulation. The combined MIEX®–coagulation evaluations were performed using a bench scale simulation of coagulation of pilot plant effluent.

A contact period of 20 minutes, a 5 percent regeneration rate, and a resin concentration of 20 mL/L were used for the MIEX® process operation during both test periods. Rock salt was used for preparing the brine for regeneration during the second test period, which resulted in reductions in bromide concentrations to less than 50 µg/L. Key test conditions for the respective test aliquots are summarized in Tables 5.14 and 5.15, and the resulting THM and HAA5 formations are shown in Tables 5.16 through 5.19. THM and HAA formation were evaluated with all samples incubated at a temperature of 20 °C and a pH of 8.0.

Table 5.14
Treatment conditions at HWTP during Week 1

	MIEX® Concentration (mL/L)	Contact Time (minutes)	MIEX® Regeneration (percent)	Alum Dose (mg/L)	Coagulation pH	DOC (mg/L)
Raw	-	-	-	-	-	6.0
MIEX® only	20	20	5	-	-	1.9
Alum only	20	20	5	55	6.1	2.4
MIEX® plus alum	20	20	5	30	6.3	1.5

Table 5.15
Treatment conditions at HWTP during Week 4

	MIEX® Concentration (mL/L)	Contact Time (minutes)	MIEX® Regeneration (percent)	Alum Dose (mg/L)	Coagulation pH	DOC (mg/L)
Raw	-	-	-	-	-	4.3
MIEX® only	20	20	5	-	-	2.5
Alum only	20	20	5	55	6.2	2.23
MIEX® plus 10 mg/L alum	20	20	5	10	7.3	1.1
MIEX® plus 30 mg/L alum	20	20	5	30	7.2	0.5<

Table 5.16
THM concentrations from DBP formation potential experiments at HWTP – Week 1

	Chlorine Residual (mg/L)			Cl_3CH (µg/L)			$BrCl_2CH$ (µg/L)			Br_2ClCH (µg/L)			Br_3CH (µg/L)			Total THM (µg/L)		
	2 hour	1 day	3 day	2 hour	1 day	3 day	2 hour	1 day	3 day	2 hour	1 day	3 day	2 hour	1 day	3 day	2 hour	1 day	3 day
Raw water	4.7	1.5	0.8	127.1	302.0	415.0	37.4	64.0	78.0	3.1	5.6	6.5	<2	<2	<2	167.6	371.6	499.5
MIEX® treated	3.5	1.96	1.0	14.0	64.8	86.7	15.8	36.6	47.5	11.9	21.0	25.0	<2	2.0	2.2	41.7	124.4	161.4
Alum only	3.45	1.9	1.1	22.7	71.7	115.4	22.2	45.8	55.0	11.6	20.1	22.0	<2	<2	<2	56.5	137.6	192.4
MIEX® + Alum	2.9	1.9	1.4	5.4	29.2	53.4	11.1	28.3	38.4	10.7	22.4	26.1	2.5	3.7	4.1	29.7	83.6	122.0

Table 5.17
THM concentrations from DBP formation potential experiments at HWTP – Week 4

	Chlorine residual (mg/L)			Cl_3CH (µg/L)			$BrCl_2CH$ (µg/L)			Br_2ClCH (µg/L)			Br_3CH (µg/L)			Total THM (µg/L)		
	2 hour	1 day	3 day	2 hour	1 day	3 day	2 hour	1 day	3 day	2 hour	1 day	3 day	2 hour	1 day	3 day	2 hour	1 day	3 day
Raw water	6.5	3.0	1.7	102.0	242.1	323.0	35.0	60.2	69.6	4.1	6.8	7.8	<2	<2	<2	141.1	309.1	400.4
Post-MIEX®	2.85	1.4	0.5	27.1	75.5	120.5	18.7	38.2	48.6	7.5	12.8	15.0	<2	<2	<2	53.3	126.5	184.1
Alum only	3.3	1.9	1.5	20.4	56.4	89.9	19.0	37.3	47.6	10.9	17.3	20.1	<2	<2	<2	50.3	111	157.6
Post-MIEX® + 10 mg/L alum	3.1	2.2	1.7	6.9	24.7	47.5	7.4	18.8	26.2	6.3	12.8	16.4	<2	2.2	2.5	20.6	58.5	92.6
Post - MIEX® + 30 mg/L alum	3.0	2.0	1.8	6.2	25.0	42.6	7.7	19.4	26.3	6.6	13.0	16.2	<2	2.0	2.3	20.5	59.4	87.4

Table 5.18
HAA5 concentrations from DBP formation potential experiments at HWTP – Week 1

	ClAA (μg/L)			BrAA (μg/L)			Cl_2AA (μg/L)			Br_2AA (μg/L)			Cl_3AA (μg/L)			Total HAA5 (μg/L)		
	2 hour	1 day	3 day	2 hour	1 day	3 day	2 hour	1 day	3 day	2 hour	1 day	3 day	2 hour	1 day	3 day	2 hour	1 day	3 day
Raw water	6.3	11.1	8.2	<2	<2	<2	71.5	107.3	121.4	<2	<2	<2	114.7	198.9	225.0	192.5	317.3	354.6
Post-MIEX®	4.9	7.3	5.1	<2	<2	<2	12.1	22.7	32.0	2.4	3.4	4.0	11.3	24.3	35.3	30.7	57.7	76.4
Alum only	7.2	7.4	4.8	<2	<2	<2	13.2	24.6	33.1	<2	2.6	3.1	11.5	28.0	37.5	31.9	62.6	78.5
Post-MIEX® + alum	10.3	8.3	3.4	<2	<2	<2	5.0	9.4	13.9	2.3	3.5	4.5	3.6	7.9	12.0	21.2	29.1	33.8

Table 5.19
HAA5 concentrations from DBP formation potential experiments at HWTP – Week 4

	ClAA (μg/L)			BrAA (μg/L)			Cl_2AA (μg/L)			Br_2AA (μg/L)			Cl_3AA (μg/L)			Total HAA5 (μg/L)		
	2 hour	1 day	3 day	2 hour	1 day	3 day	2 hour	1 day	3 day	2 hour	1 day	3 day	2 hour	1 day	3 day	2 hour	1 day	3 day
Raw water	4.7	8.2	9.6	<2	<2	<2	51.5	83.4	97.0	1.7	1.8	1.7	102.2	201.6	230.0	160.1	295.0	338.3
Post-MIEX®	<2	3.0	3.9	<2	<2	<2	16.0	29.2	37.3	2.3	2.5	2.7	20.2	42.7	54.7	38.5	77.4	98.6
Alum only	2.7	3.0	5.5	<2	<2	<2	9.9	14.0	24.7	2.9	2.1	4.0	13.0	14.7	36.4	28.5	33.8	70.6
Post-MIEX® + 10 mg/L alum	<2	2.6	3.9	<2	<2	<2	4.0	7.7	12.1	2.2	3.2	3.6	5.0	9.7	14.1	11.2	23.2	33.7
Post-MIEX® + 30 mg/L alum	9.2	11.1	7.7	<2	<2	<2	4.2	7.8	11.3	2.6	3.6	4.0	4.8	8.9	13.3	20.8	31.4	36.3

Testing results indicate that both MIEX® alone and alum alone can significantly reduce DOC concentrations in water, and thus, the formation of THMs and HAA5. Significant additional reductions could be achieved by combining MIEX® and alum coagulation at this site and these is a potential for an overall benefit of operating in this sequence.

Although large reductions in THM and HAA formation occurred in the MIEX®–alum coagulation sequence, THM formation remained high relative to regulated levels during the first period of testing. Somewhat better reductions were achieved during the second testing period. More DOC was removed by the combined sequence in the second period, and some reduction in bromide concentrations was achieved. If additional reductions are desirable at HWTP, they might be achieved by further adjustments within the treatment train, such as use of an alternative coagulant or a higher regeneration percentage in the MIEX® process. This interrelation between the MIEX® process and coagulation emphasizes the importance of assessing the MIEX® process as part of a process sequence, rather than as a discrete process.

Analyses of Resin Activity and Buildup of DOC in the Regenerant Brine

Testing was performed to compare the effectiveness of virgin resin, regenerated resin collected at startup of the pilot plant, and regenerated resin collected at the end of the four-week period of pilot plant operation (including the initial week of start-up operation). The testing was conducted using parallel dose-response tests to assess the uptake of UV254 absorbance and DOC from a common raw water sample collected at the plant site. The concentration of DOC in this water was 5.1 mg/L and its UV254 absorbance was 0.196 cm^{-1}. Results of the dose-response tests, summarized in Table 5.20, show good agreement between the resin samples in terms of both DOC and UV254 uptake at each MIEX® dose used, and indicate that there was very little change in the activity of the resin for NOM removal over the course of the testing.

Table 5.20
Evaluation of resin activity at HWTP

Utility	MIEX® dose (mL/L)	UV254 absorbance reduction (%)	DOC removal (%)
Virgin resin	2	71	58
	4	92	72
	6	98	80
Regenerated resin at beginning of operation (one week of start-up)	2	77	60
	4	94	67
	6	98	72
Regenerated resin after five weeks of operation (four weeks of operation)	2	72	57
	4	90	70
	6	93	78

Condition of the regenerant solution was also assessed. The brine was initially prepared to a salt concentration of 125 g/L and was recovered at the end of each regeneration cycle throughout the week as was the case at the other test sites. Salt was also added to the brine to make up for any chloride lost in the MIEX® treatment sequence. Build-up of NOM as a consequence of this mode of operation was assessed by sampling and analyzing the DOC level in the brine at the end of each week before preparing a fresh batch of regenerant. The results of these analyses are listed in Table 5.21. The DOC concentrations are higher than those at the other test sites, but lower than the 8 mg/L encountered during operation at the full-scale WGTP.

Table 5.21
DOC buildup in the regenerant brine at HWTP

Week of pilot operation	DOC concentration at end of week of operation (g/L)
1	2.6
2	2.9
3	3.9
4	3.6

Fort Thomas Water Treatment Plant

The FTWTP is owned and operated by the Northern Kentucky Water District and treats water from the Ohio River. During testing, a 1 mg/L dose of potassium permanganate ($KMnO_4$) was added at the raw water intake ahead of the pilot plant. The $KMnO_4$ contact time prior to reaching the pilot plant was several minutes and there was a potential for impact on the MIEX® process as a consequence; however, this was not evaluated. DOC concentrations in the raw water were relatively low, but still at levels that result in significant formation of THMs and HAAs. The focus of study at this location was on the effect of the MIEX® process on THM and HAA formation.

The FTWTP employs a conventional coagulation, flocculation, sedimentation, and filtration sequence, and the effect of MIEX® pretreatment on coagulant performance is a consideration. Table 5.22 summarizes the raw water quality during the bench and pilot testing conducted between May 1 and June 7, 2003.

Table 5.22
Fort Thomas WTP raw water quality

	Average	Minimum	Maximum
Temperature (°C)	17	15	20
pH	7.78	7.63	7.94
Turbidity (ntu)	78.21	26.20	158.00
UV254 (cm^{-1})	0.082	0.063	0.120
DOC (mg/L)	1.79	1.19	2.46
SUVA	4.6	NA	NA
Total alkalinity (mg/L)	66	60	73
Total hardness (mg/L)	113	94	121
Bromide (mg/L)	0.04	0.03	0.05
Sulfate (mg/L)	73.7	50.5	97.7
Chloride (mg/L)	17.4	13.5	19.5
TDS (mg/L)	190	161	205

The testing period coincided with wet weather conditions that resulted in high turbidity in the raw water throughout the entire period. TDS concentrations were generally somewhat higher than at the HWTP and lower than at the MWTP. However, the average sulfate concentration was 73.7 mg/L, which was significantly higher than the concentrations at either HWTP or MWTP. This is significant because the divalent charge of the sulfate ion heightens its potential for competition with other ions for ion exchange sites. The alkalinity was moderately low, while hardness was higher than at HWTP, but lower than at MWTP.

Bench-scale Testing

The results of bench scale MIEX® dose-response tests for reduction or removal of UV254 absorbance and DOC are shown on Figures 5.20 and 5.21 in terms of percent removal. Similar to HWTP, the results indicated that longer detention time and a higher MIEX® resin concentration had a significant impact on the reduction of UV254 and DOC. This could have been affected by differences in raw water characteristics that included differences in sulfate levels. However, the raw water at the FTWTP also appeared to contain a higher percentage of organic material that could not be readily removed than the waters treated at HWTP and MWTP. Unlike the testing at the HWTP, no bromide removal was observed in the bench scale test results shown in Table 5.23.

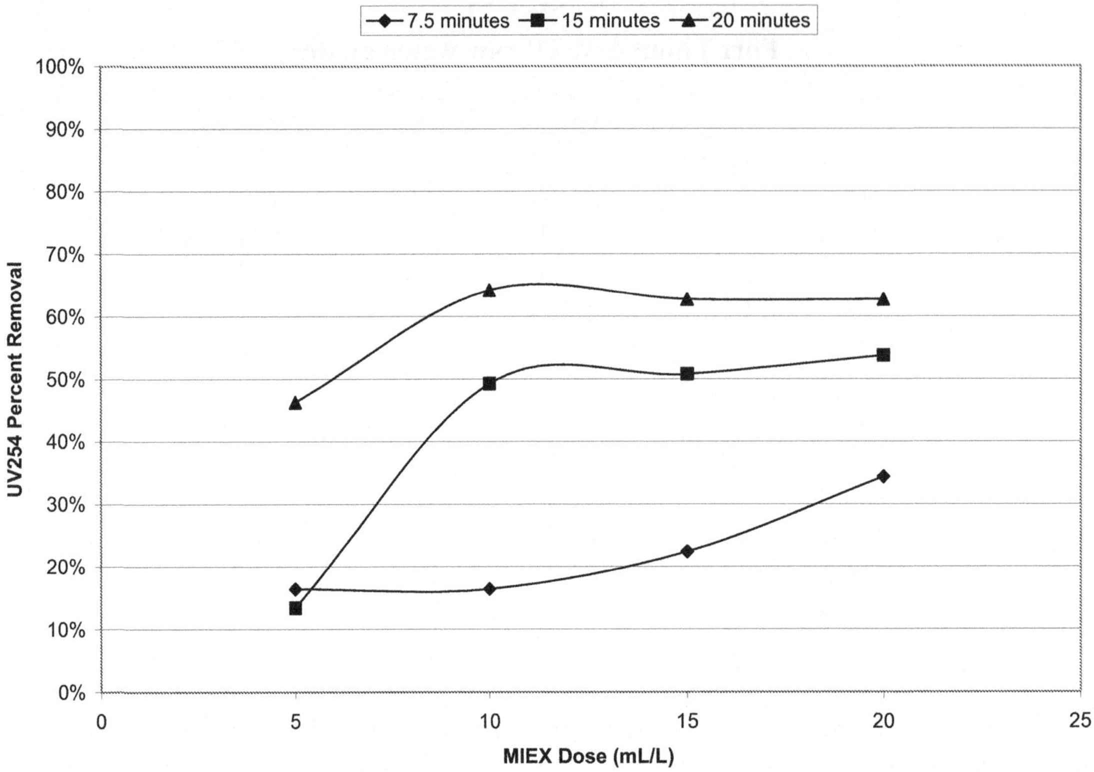

Figure 5.20 Percent UV254 absorbance reduction during dose-response testing at FTWTP

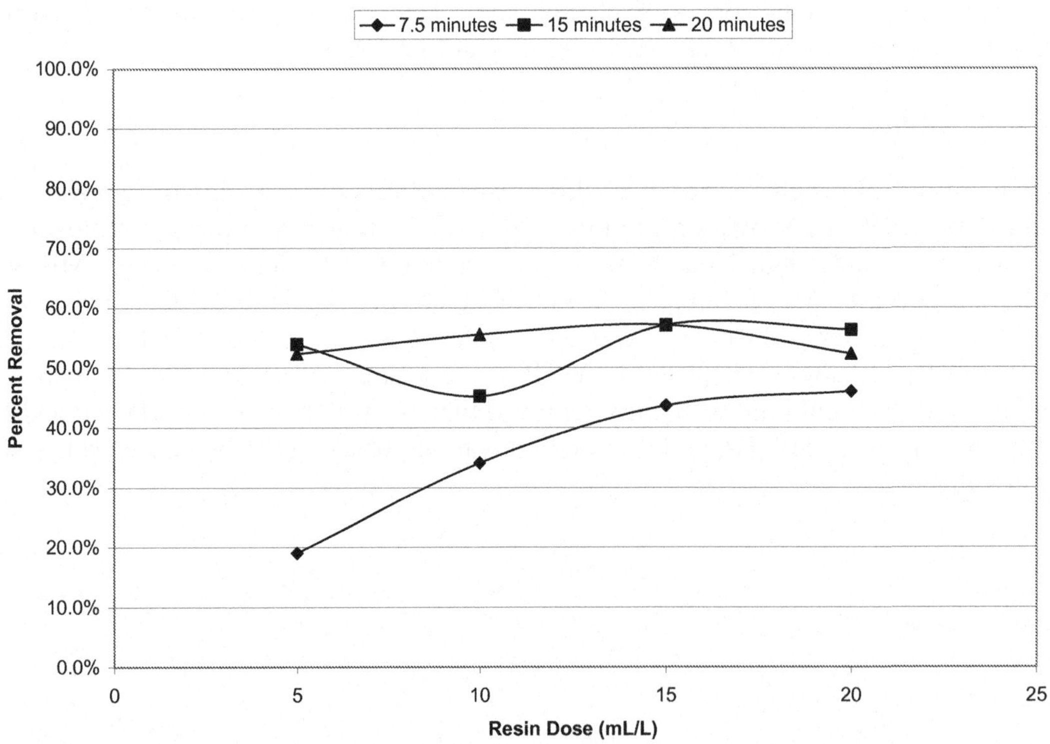

Figure 5.21 Percent DOC removal during dose-response testing at FTWTP

Table 5.23
Bromide analysis results from dose-response tests at FTWTP (mg/L)

95 percent preloaded resin and 5 percent fresh resin	7.5 minutes	15 minutes	20 minutes
Raw water	0.034	0.034	0.034
5 mL/L	0.036	0.037	0.037
10 mL/L	0.036	0.038	0.037
15 mL/L	0.036	0.038	0.038
20 mL/L	0.038	0.038	0.037

Pilot Testing

Based on the results of the dose-response testing, pilot testing was performed at resin concentrations of 15 and 20 mL/L, regeneration rates up to 10 percent, and a contact time of 15 minutes. Results of the bench scale test indicated some benefit in terms of added reduction of UV254 absorbance at a contact time of 20 minutes, but these did not appear to be significant in terms of added removal of DOC. Specific operating conditions during the four consecutive weeks of pilot testing at the Fort Thomas WTP are summarized below:

- Week 1 - 20 mL/L resin, 15 minute contact time, and 5 percent of flow sent to regeneration.
- Week 2 - 15 mL/L resin, 15 minute contact time, and 5 percent of flow sent to regeneration.
- Week 3 - 15 mL/L resin, 15 minute contact time, and 10 percent of flow sent to regeneration.
- Week 4 - 15 mL/L resin, 15 minute contact time, and 5 percent of flow sent to regeneration.

Results of UV254 absorbance and DOC reduction/removal testing are shown on Figures 5.22 through 5.25. These removals are not as high as those achieved at MWTP and HWTP, but are similar to those achieved during the bench-scale dose-response testing at this location. No discernible trend in removals was observed when using different resin doses or regeneration conditions, which may indicate that there is little opportunity for additional removal of NOM from this water once a certain level is reached. This observation is consistent with those of the bench scale results shown in Figures 5.20 and 5.21.

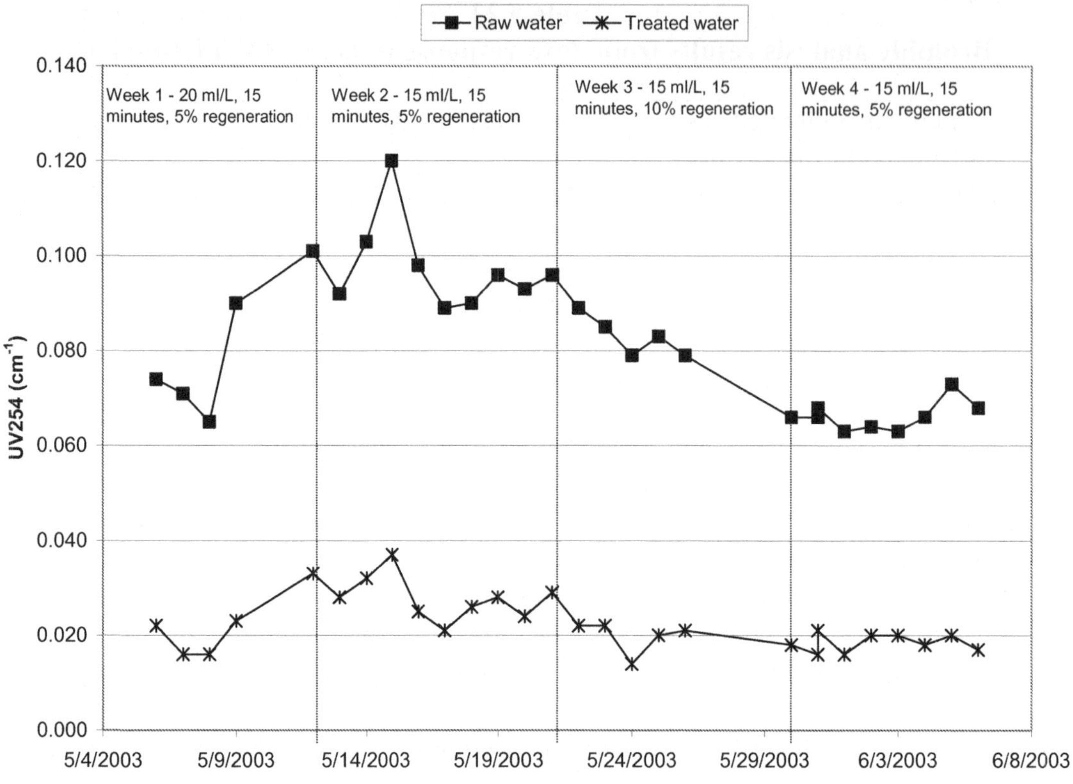

Figure 5.22 UV254 absorbance reduction during pilot plant operation at FTWTP

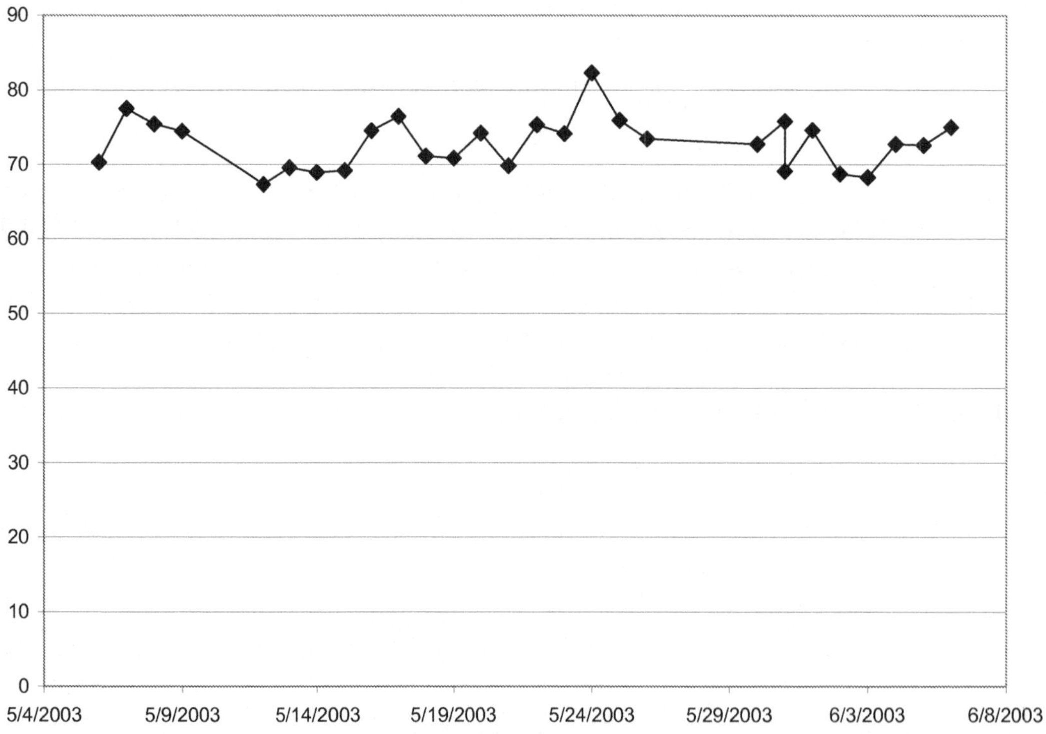

Figure 5.23 Percent UV254 absorbance reduction during pilot plant operation at FTWTP

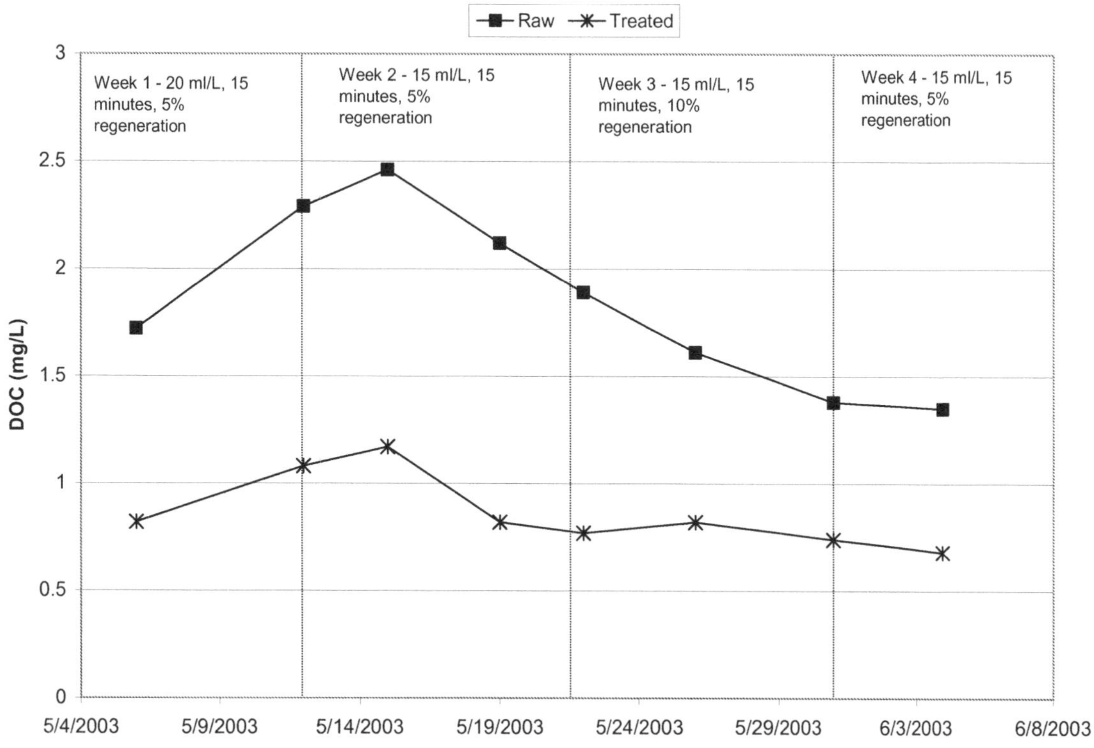

Figure 5.24 DOC removal during pilot plant operation at FTWTP

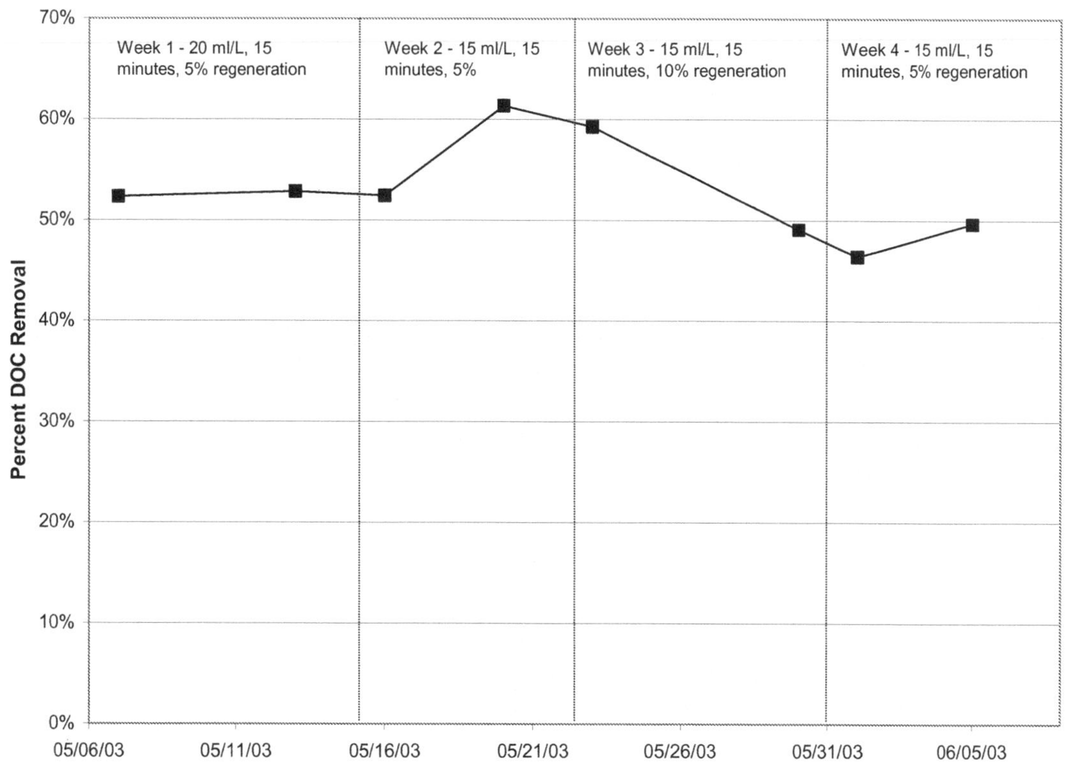

Figure 5.25 Percent DOC removal during pilot plant operation at FTWTP

Bromide was not removed during this testing and there were times when the data indicate raw water levels greater than the levels in the MIEX treated water, as shown on Figure 5.26. Solar salt was used to prepare the brine at this location and the bromide in this salt may have impeded the removal of bromide, as appeared to be the case at the HWTP when solar salt was used. In addition, sulfate levels are higher at this location and this may have been a factor that impeded uptake.

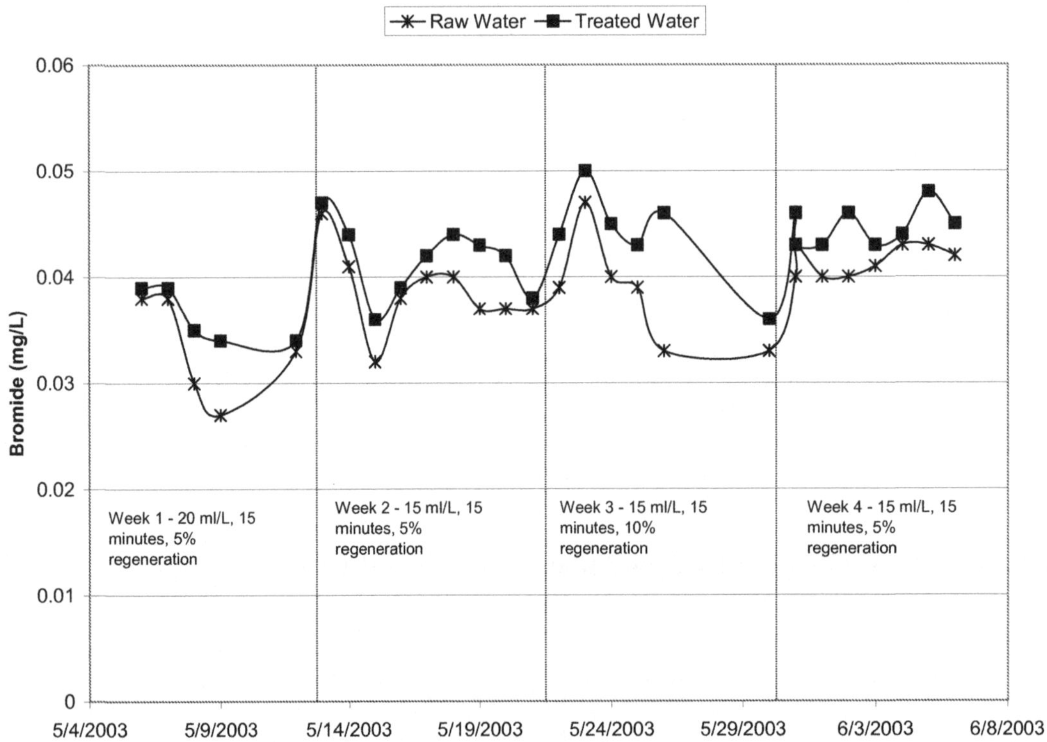

Figure 5.26 Bromide concentrations during pilot plant operation at FTWTP

Detention Time Study

Detention time studies were performed each week to determine the effect of reducing the contact time to one half of the normal detention time. This change was implemented by bypassing one of the two contactors to reduce the contacting volume by half. A period of two hours was allowed for the change to take place across the pilot plant. The results of these studies are shown in Table 5.24.

Table 5.24
Detention time study results from the FTWTP

	Week 1 - 20 mL/L 5% regeneration		Week 2 - 15 mL/L 5% regeneration		Week 3 - 15 mL/L 10% regeneration		Week 4 - 15 mL/L 5% regeneration	
	Percent removal 7.5 minutes	Percent removal 15 minutes	Percent removal 7.5 minutes	Percent removal 15 minutes	Percent removal 7.5 minutes	Percent removal 15 minutes	Percent removal 7.5 minutes	Percent removal 15 Minutes
UV254 (cm^{-1})	62	72	65	72	52	76	63	72
DOC (mg/L)	42	53	52	61	44	49	39	48

Effect of MIEX® Process on General Water Quality Characteristics

The effect of MIEX® treatment on general water quality characteristics is shown in Table 5.25.

Table 5.25
Average concentrations of water quality constituents at FTWTP after MIEX® treatment

	pH	Total alkalinity (mg/L)	Total hardness (mg/L)	Sulfate (mg/L)	Chloride (mg/L)	TDS (mg/L)
Raw	7.78	66	113	73.7	17.4	190
Week 1	7.80	69	121	60	60	211
Week 2	7.89	63	111	54	39	186
Week 3	7.74	63	103	42	49	214
Week 4	7.91	67	115	54	54	204

A slight reduction in sulfate concentration was observed as expected. Chloride concentrations increased as a result of exchange for anions taken up by the resin. TDS concentrations did not change significantly, possibly due to a balance between the anions removed by the resin and the chloride being released. Little change was observed in the concentrations of hardness and alkalinity, and in pH. Figures 5.27 through 5.29 indicate the trends for sulfate, chloride, and TDS for the four weeks of pilot testing. The increase in chloride concentration was in the range of 20 to 30 mg/L.

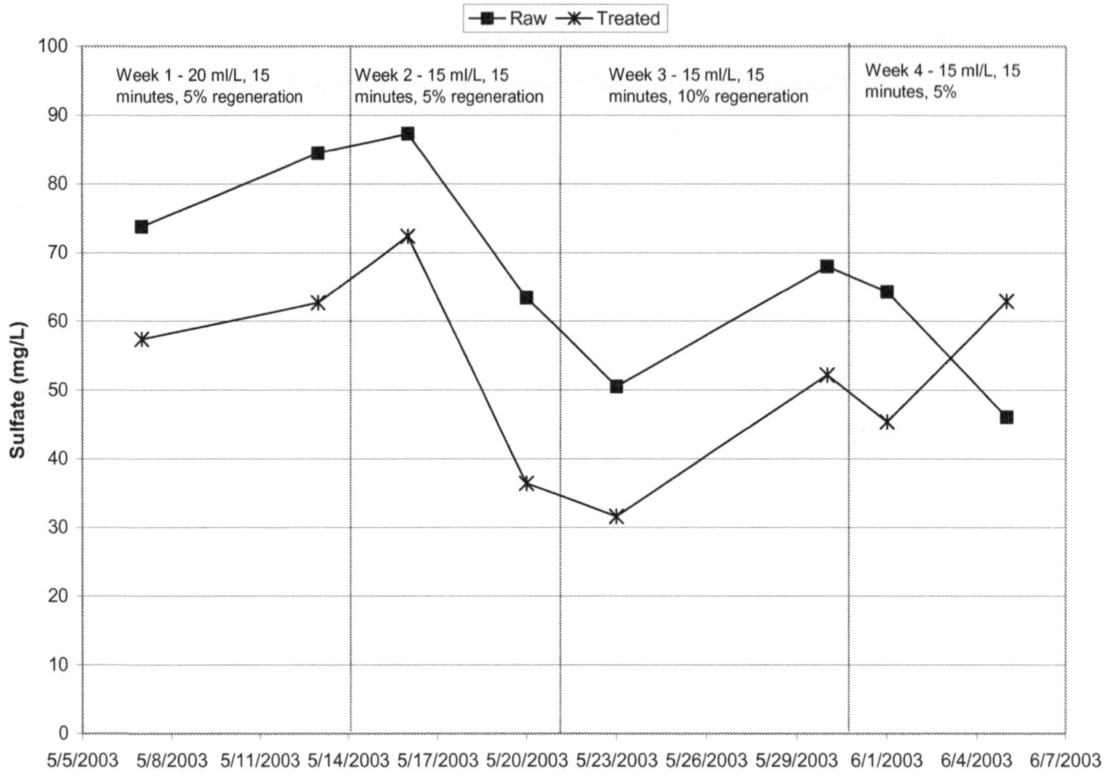

Figure 5.27 Sulfate concentrations during pilot plant operation at FTWTP

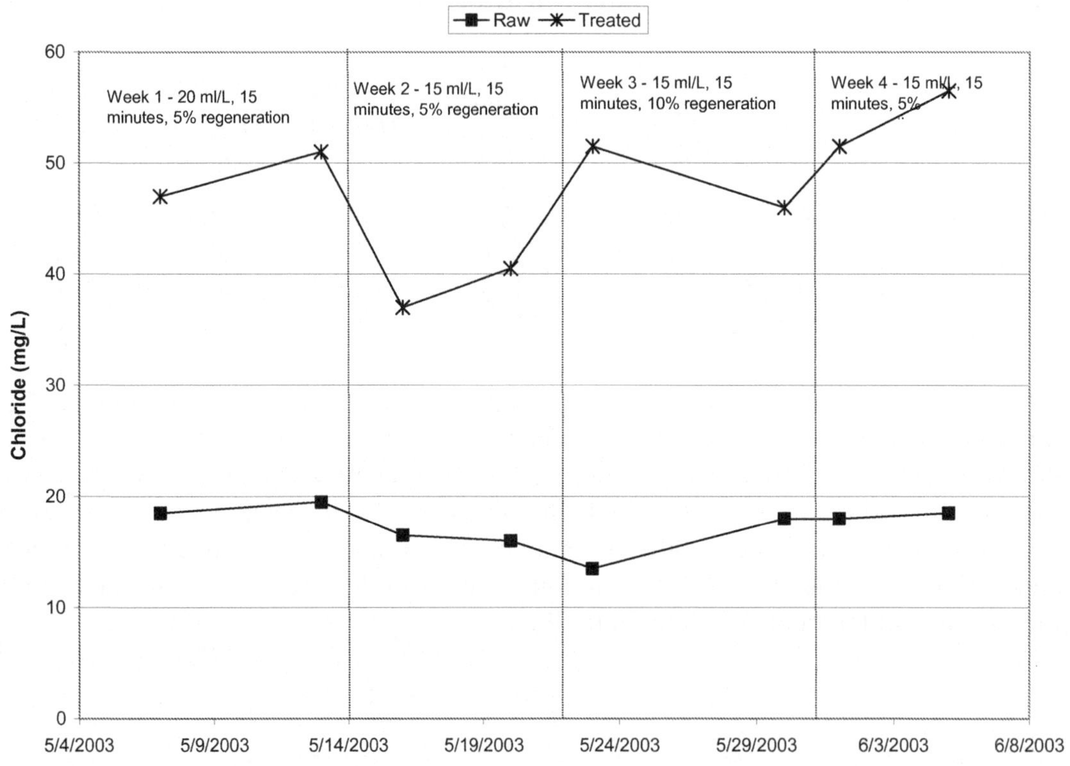

Figure 5.28 Chloride concentrations during pilot plant operation at FTWTP

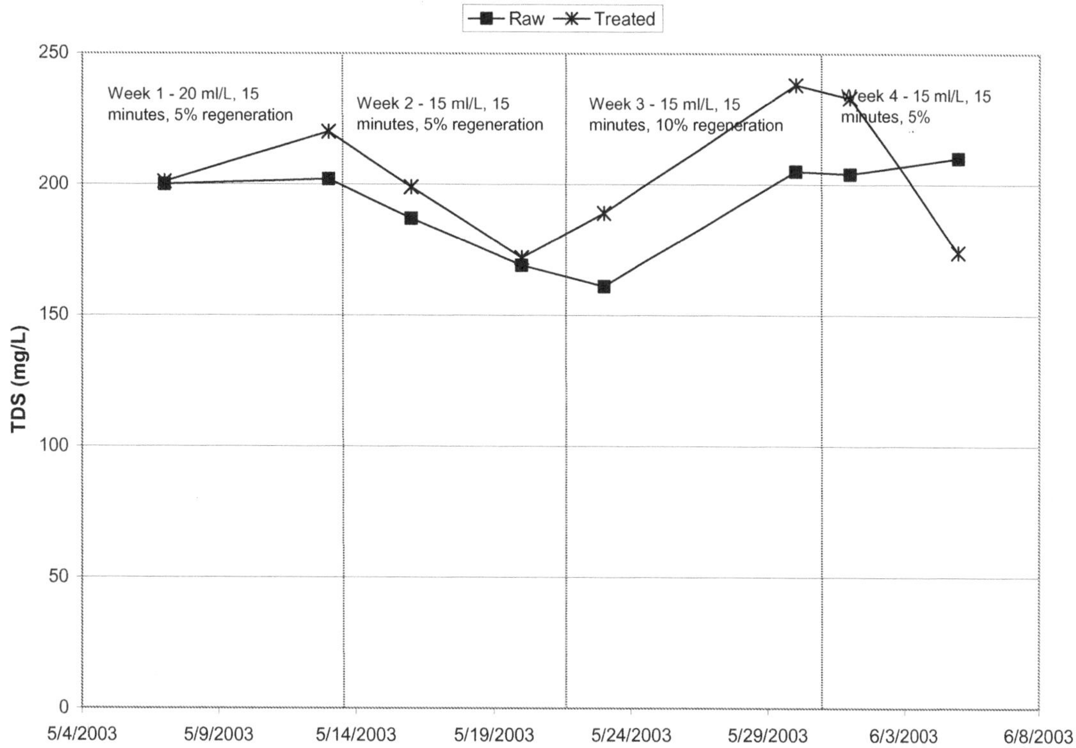

Figure 5.29 TDS concentrations during pilot plant operation at FTWTP

Coagulation Study

Coagulation studies were conducted at FTWTP during the initial startup week in a manner similar to those at HWTP. Treated water from the pilot plant was used in the bench scale coagulation testing at a constant temperature that matched the temperature in the full-scale plant and results were compared with a control set of coagulation tests that used raw water without MIEX® pretreatment.

Similar testing was performed for a second set of coagulant assessments at the end of the pilot study. MIEX® process operating conditions for the first and second sets of tests were similar, except that the resin concentration was 20 mL/L during the first test period and 15 mL/L during the second. A detention time of 15 minutes and a regeneration rate of 5 percent were used during both test periods. The MIEX®-coagulation process sequence was tested in parallel with a control condition using coagulated water with no MIEX® pretreatment.

Coagulation at the Fort Thomas WTP involves a blend of two coagulants; ferric sulfate at an average dosage of 5 mg/L, and a commercial acidified alum-polymer blend at a dose that is varied in response to changes in raw water quality. The role of the ferric sulfate in this blended coagulant scheme is to provide an improvement in the removal of NOM.

The concept of maintaining a constant ferric sulfate dose while varying the dose of the acidified alum was applied during the coagulant testing, and 5 mg/L of ferric sulfate was applied for each coagulation condition that was tested. The acidified alum dosage was varied and the pH of the water was also adjusted using sodium hydroxide or sulfuric acid as required to provide a matrix of test conditions.

Figures 5.30 and 5.31 present results from the initial set of these coagulation evaluations. These results show different responses between the MIEX®–coagulation sequence and the use of coagulation only in that the coagulation only approach worked best at higher pH levels. It may also be possible that lower acidified alum doses could be applied for the coagulation only approach. No contour plot was developed for UV254 absorbance for the MIEX®–coagulation sequence, as the levels were uniformly reduced to a very narrow range of 0.013 to 0.016 cm^{-1}, which does not provide adequate variation for a meaningful plot of trends because the analytical accuracy of the UV254 analysis results in scatter that can comprise a significant component of the difference at these low levels. However, it can be seen that these levels were uniformly lower than those achieved by coagulation alone as shown on Figure 5.32.

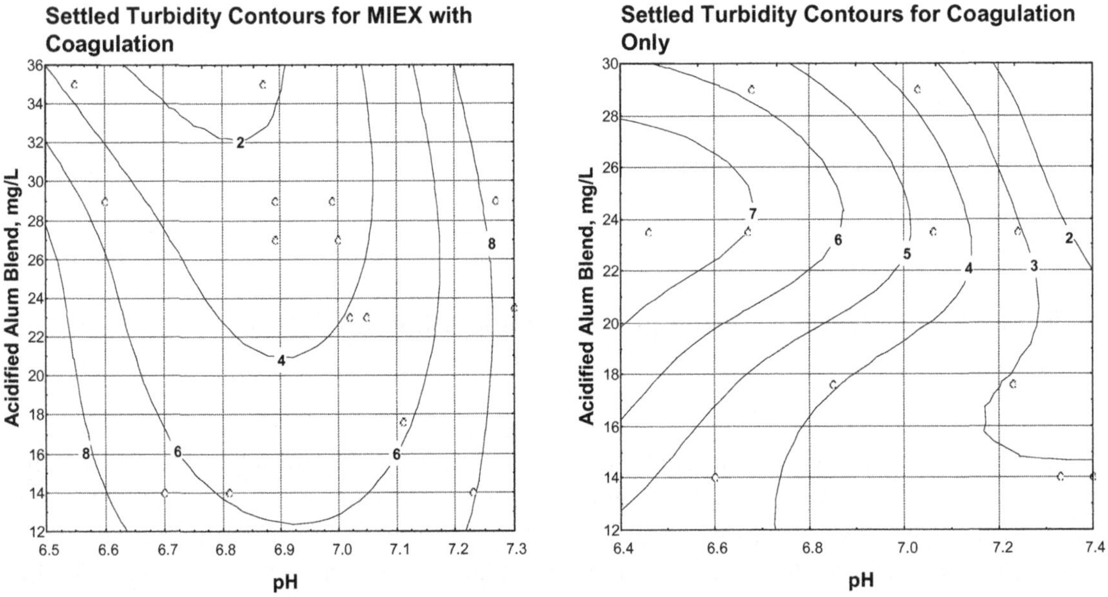

Figure 5.30 Settled turbidity from coagulation evaluations during Week 1 at FTWTP

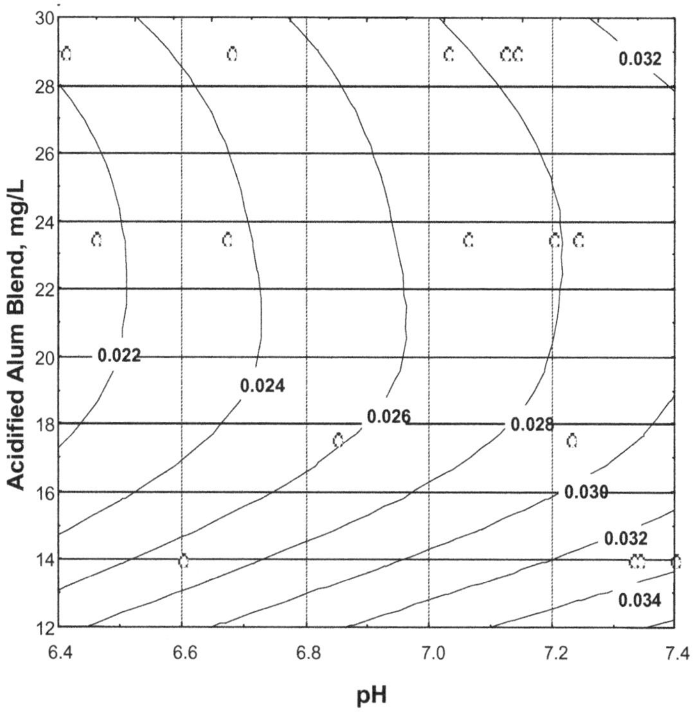

Figure 5.31 Settled water UV254 absorbance from coagulation evaluations without MIEX® pretreatment during Week 1 at FTWTP

The results of the second set of coagulation tests (Week 4) differed somewhat from Week 1 results, but the general trends were similar, indicating that an optimum acidified alum dose for turbidity removal can be identified at a slightly lower dose for the coagulation only approach than for the MIEX® – coagulation sequence, as shown on Figure 5.32. UV254 results were similar to those in the initial set of coagulation tests in that a uniform reduction was achieved to levels less than 0.015 cm^{-1} for the MIEX®-acidified alum sequence, while the variation in pH and acidified alum dose had a greater effect on UV254 in the acidified alum-only testing, shown on Figure 5.33. UV254 results for the MIEX – coagulation approach were also uniformly lower than the coagulation only approach as was the case for the first set of tests.

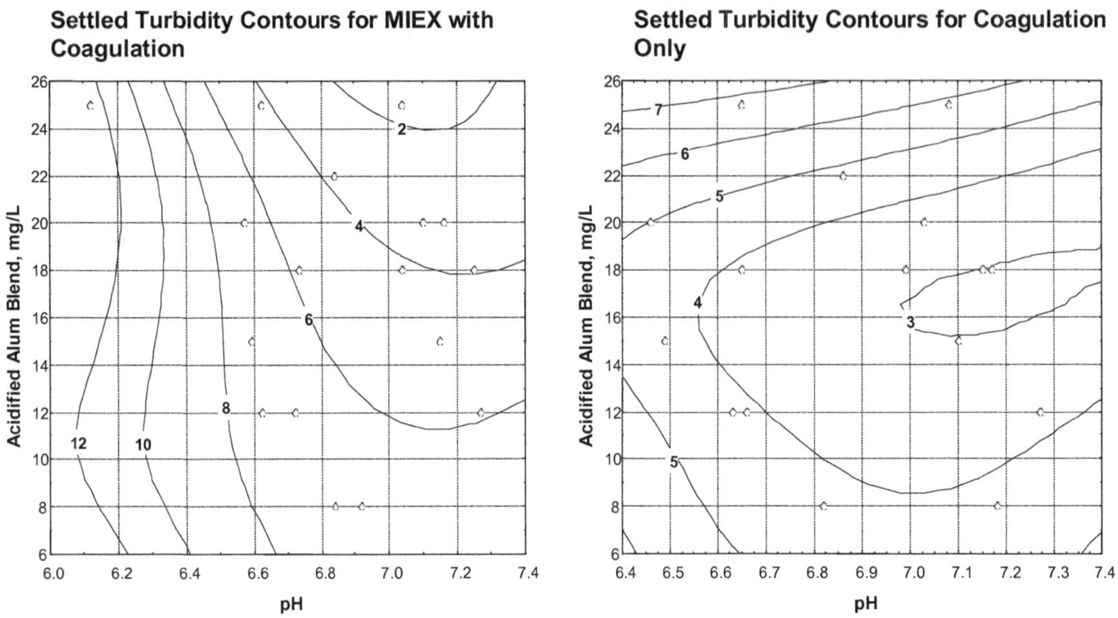

Figure 5.32 Settled turbidity from coagulation evaluations during Week 4 at FTWTP

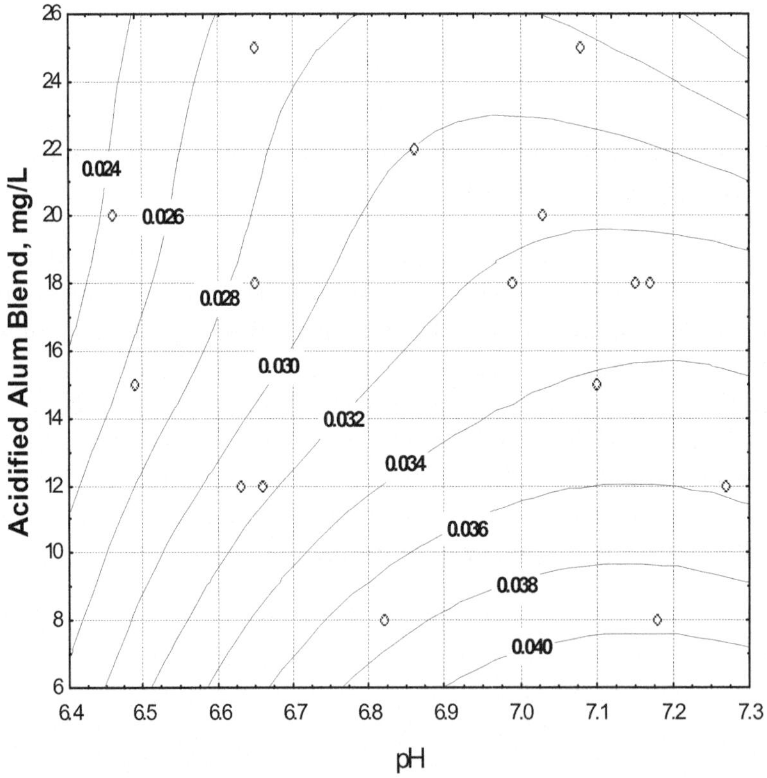

Figure 5.33 Settled water UV254 absorbance from coagulation evaluations without MIEX® pretreatment during Week 4 at FTWTP

Testing was also performed during the second set of tests to assess the effect of MIEX® resin on coagulation if only one coagulant was applied. This testing was performed with ferric sulfate, which had been shown during prior testing to achieve significant turbidity reduction when used alone with MIEX® pretreatment. Results of this testing confirmed the applicability of ferric sulfate with MIEX® pretreatment to produce settled turbidity levels of 2 ntu or less (one indication of potential for effective coagulation for raw water with a turbidity greater than 10 ntu) at doses of 10 mg/L or slightly less as shown on Figure 5.34, This figure also indicates that coagulation with ferric sulfate in the absence of MIEX® pretreatment was not effective for achieving the target settled turbidity level at this dosage . These results indicate that MIEX® pretreatment is beneficial for coagulation when using a single coagulant and is consistent with the observations made at HWTP where a single coagulant is used. The situation is more complex for the multiple coagulant approach that is presently applied at the FTWTP.

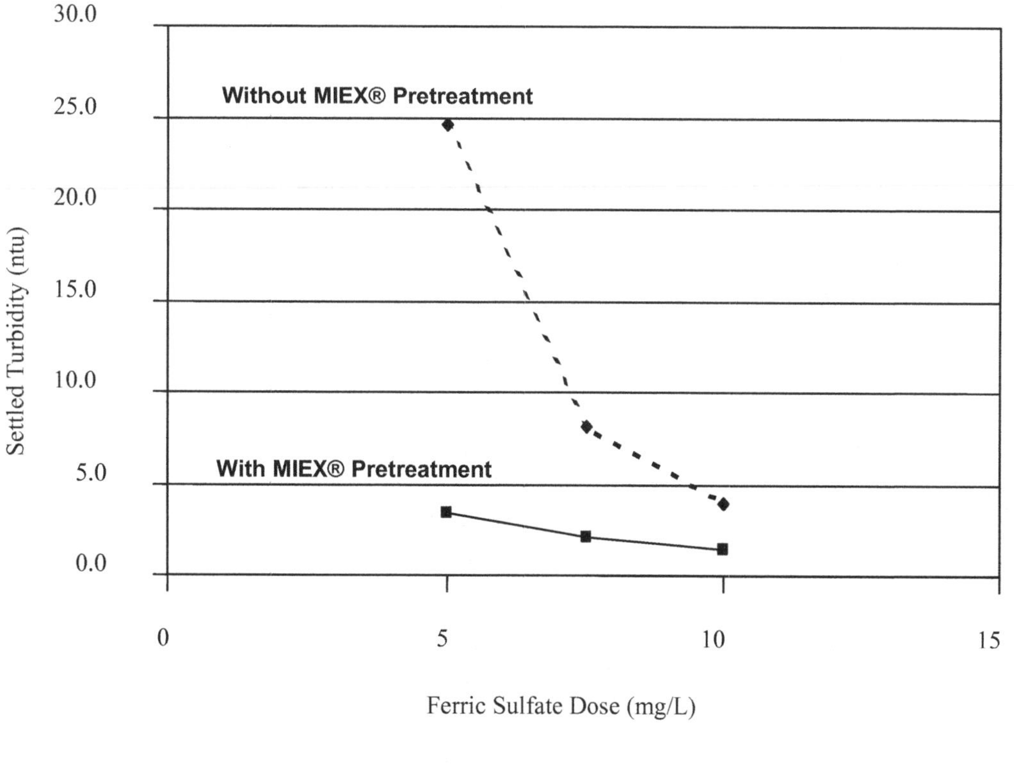

Figure 5.34 Effect of ferric sulfate alone on coagulation following MIEX® at FTWTP during Week 4

Analyses of THM and HAA5 Formation

The effects of NOM removal by the MIEX® process and a MIEX®–coagulation sequence on the formation of THMs and HAA5 were evaluated during two testing periods. Coagulation characterizations described in the preceding section were used to select the alum doses for the coagulation-only and MIEX®– coagulation alternatives. Disinfection by-product formation was characterized under UFC that have been described, with the samples incubated at 20 °C and pH 8.0.

MIEX® treated water for this testing was obtained from the pilot plant, and the combined MIEX®–coagulation evaluations were performed using a bench scale coagulation simulation of pilot plant effluent. A contact period of 15 minutes, a regeneration rate of 5 percent, and a resin concentration of 15 mL/L were used for the MIEX® process operation during both test periods. The combination of acidified alum and ferric sulfate used in the full-scale plant was used in both the MIEX®-coagulation and the coagulation-only simulations in the first period. However, the effectiveness of ferric sulfate as the sole coagulant when used in combination with MIEX® pretreatment resulted in use of this approach for the MIEX®–coagulation simulation during the second test period. Key testing conditions are summarized in Tables 5.26 and 5.27 for the two testing periods, and the resulting THM and HAA5 concentrations are shown in Tables 5.28 through 5.31.

Table 5.26
Treatment conditions at FTWTP during Week 1

	MIEX® concentration (mL/L)	Contact time (minutes)	MIEX® regeneration rate (percent)	Acidified alum dose (mg/L)	Ferric sulfate (mg/L)	pH	DOC (mg/L)
Raw	-	-	-	-	-	7.9	1.5
MIEX® only	20	15	5	-	-	7.9	1.3
Ferric sulfate - acidified alum only	-	15	5	23.5	5	7.1	0.6
MIEX® plus ferric sulfate - acidified alum	20	15	5	23.5	5	7.0	0.5

Table 5.27
Treatment conditions at FTWTP during Week 4

	MIEX® concentration (mL/L)	Contact time (minutes)	MIEX® regeneration rate (percent)	Acidified alum dose (mg/L)	Ferric sulfate (mg/L)	pH	DOC (mg/L)
Raw	-	-	-	-	-	7.8	1.6
MIEX® only	15	15	5	-	-	7.7	0.8
Ferric sulfate - acidified alum only	-	15	5	18	5	7.2	0.9
MIEX® plus ferric sulfate	15	15	5	-	10	7.3	1.0

Table 5.28
THM concentrations from DBP formation potential experiments at FTWTP – Week 1

	Chlorine Residual (mg/L)			Cl_3CH (µg/L)			$BrCl_2CH$ (µg/L)			Br_2ClCH (µg/L)			Br_3CH (µg/L)			Total THM (µg/L)		
	2 hour	1 day	3 day	2 hour	1 day	3 day	2 hour	1 day	3 day	2 hour	1 day	3 day	2 hour	1 day	3 day	2 hour	1 day	3 day
Raw Water	4.3	1.7	0.5	51.6	134.9	184.9	15.0	26.3	30.7	1.8	2.7	3.0	<.2	<2	<.2	68.4	163.9	218.6
MIEX® only	2.2	1.9	1.4	10.7	36.2	51.8	6.5	15.1	19.9	3.4	6.5	7.8	<2	<2	<2	20.6	57.8	79.5
Coag. only	2.9	1.4	0.8	24.8	61.1	86.4	10.9	19.0	24.5	2.6	4.2	5.1	<2	<2	<2	38.3	84.3	116.0
MIEX® + Coag			1.6			41.0			18.3			7.7			0.9			67.9

Table 5.29
THM concentrations from DBP formation potential experiments at FTWTP – Week 4

	Chlorine residual (mg/L)			Cl_3CH (µg/L)			$BrCl_2CH$ (µg/L)			Br_2ClCH (µg/L)			Br_3CH (µg/L)			Total THM (µg/L)		
	2 hour	1 day	3 day	2 hour	1 day	3 day	2 hour	1 day	3 day	2 hour	1 day	3 day	2 hour	1 day	3 day	2 hour	1 day	3 day
Raw Water	2.15	1.63	1.01	48.1	98.3	132.2	10.7	17.0	23.3	3.3	4.6	4.8	<2	<2	<2	62.1	119.9	160.3
MIEX® only	2.35	2.12	1.50	13.8	27.0	38.6	5.1	12.4	17.4	5.6	10.1	10.9	<2	<2	<2	24.5	49.5	66.9
Coag. only	2.00	1.48	1.10	23.1	61.1	83.2	8.0	14.9	20.4	4.2	6.7	7.1	<2	<2	<2	35.3	82.7	110.7
MIEX® + Coag.	2.01	1.52	1.07	4.0	17.6	31.8	4.1	11.1	17.2	5.1	10.0	11.6	<2	<2	<2	13.2	38.7	60.6

Table 5.30
HAA5 concentrations from DBP formation potential experiments at FTWTP – Week 1

	ClAA (μg/L)			BrAA (μg/L)			Cl_2AA (μg/L)			Br_2AA (μg/L)			Cl_3AA (μg/L)			Total HAA5 (μg/L)		
	2 hour	1 day	3 day	2 hour	1 day	3 day	2 hour	1 day	3 day	2 hour	1 day	3 day	2 hour	1 day	3 day	2 hour	1 day	3 day
Raw Water	2.1	3.5	5.1	<2	<2	<2	25.0	52.3	64.5	<2	<2	<2	33.4	62.7	72.7	60.5	118.5	142.3
MIEX® only	<2	<2	<2	<2	<2	<2	4.6	12.6	17.0	<2	<2	<2	5.1	12.5	16.3	9.7	25.1	23.3
Coagulation only	<2	<2	1.5	<2	<2	<2	10.0	17.8	25.7	<2	<2	<2	12.4	19.9	26.3	22.4	37.7	53.5
MIEX® + coagulation	<2	<2	<2	<2	<2	<2	2.7	8.5	8.6	<2	<2	<2	4.9	8.7	8.3	7.6	17.2	16.9

Table 5.31
HAA5 concentrations from DBP formation potential experiments at FTWTP – Week 4

	ClAA (μg/L)			BrAA (μg/L)			Cl_2AA (μg/L)			Br_2AA (μg/L)			Cl_3AA (μg/L)			Total HAA5 (μg/L)		
	2 hour	1 day	3 day	2 hour	1 day	3 day	2 hour	1 day	3 day	2 hour	1 day	3 day	2 hour	1 day	3 day	2 hour	1 day	3 day
Raw water	<2	<2	3.1	<2	<2	<2	14.6	28.7	36.5	<2	<2	<2	18.9	34.7	41.7	33.5	63.4	81.3
MIEX® only	<2	<2	<2	<2	<2	<2	5.1	7.4	10.2	<2	<2	<2	7.2	8.0	9.6	12.3	15.4	19.8
Coag only	<2	<2	2.0	<2	<2	<2	6.5	14.8	19.9	<2	<2	<2	8.8	16.5	19.4	15.3	31.3	41.3
MIEX® + Coag	<2	<2	<2	<2	<2	<2	<2	4.6	7.2	<2	<2	<2	3.1	4.3	5.6	3.1	8.9	12.8

These results indicate that the MIEX® process alone and alum/ferric sulfate coagulation reduce DOC concentrations to similar levels and that this parameter does not provide a basis for delineating a difference between these two processes. A clearer difference was seen in the relative reductions in UV254 as described previously. Differences are also observed in the formation of THMs and HAA5s. Three-day HAA5 concentrations were significantly lower than the coagulation only approach for both the MIEX® only and the MIEX®–coagulation sequence, and were well within the regulatory limit of 60 μg/L. Three-day THM concentrations in the coagulation-only were also significantly greater than those observed in the MIEX® only and the MIEX® – coagulation sequence, again showing differences in DBP formation attributes in spite of similar DOC levels when coagulation and MIEX® are used.

Analyses of Resin Activity and Buildup of DOC in the Regenerant Brine

Testing was performed to compare the activity of virgin resin, regenerated resin collected at startup of the pilot plant, and regenerated resin collected at the end of the four-week period of pilot plant operation (including the initial week of start-up operation). Parallel dose-response tests were used to assess the uptake of UV254 and DOC from a common raw water sample. The concentration of DOC in this water was 2.0 mg/L and its UV254 absorbance was 0.096 cm^{-1}. Results of the dose-response tests, which are summarized in Table 5.32, show generally good agreement between the three resin samples in terms of both DOC and UV254 uptake for the 1 and 2 mL/L results. The DOC result at a dose of 4 mL/L of regenerated resin after four weeks of operation are outside of the trend exhibited at the other resin concentrations and may not be representative.

Table 5.32
Evaluation of resin activity at FTWTP

Utility	MIEX® dose (mL/L)	UV254 absorbance reduction (%)	DOC removal (%)
Virgin resin	1	41	36
	2	48	58
	4	66	72
Regenerated resin at the beginning of operation (one week of startup)	1	36	35
	2	43	57
	4	66	71
Regenerated resin after five weeks of operation (four weeks of operation)	1	38	39
	2	52	53
	4	55	9

Buildup of DOC as a consequence of re-use of the brine was a consideration, and was assessed by analyzing the brine for DOC concentration at the end of each week before preparing fresh regenerant batch. The results of these analyses are listed in Table 5.33. These levels are relatively low, and reflect the low DOC content of the raw water.

Table 5.33
DOC buildup in the regenerant brine at FTWTP

Week of pilot operation	DOC concentration at end of week of operation (g/L)
1	0.6
2	1.4
3	1.5
4	1.4

Alfred Merritt Smith Water Treatment Facility

The Alfred Merritt Smith Water Treatment Facility (AMSWTF) is a 600 mgd ozonation and direct filtration facility in Boulder City, NV, owned and operated by the Southern Nevada Water Authority and treats water from Lake Mead. Lake Mead is supplied by the Colorado River, by surface runoff from the Las Vegas Wash, and by the Muddy River. Perchlorate in the water from the Las Vegas Wash, and arsenic in the Muddy River water have the potential to affect raw water quality in the lake. The MIEX® process was evaluated to assess the effects on DOC, formation of THMs and HAA5, and bromide, arsenic and perchlorate.

Testing at this site was performed during the period from July 10 to August 13, 2003. Table 5.34 summarizes the raw water quality during bench scale and pilot scale testing.

Table 5.34
AMSWTF raw water quality

	Average	Minimum	Maximum
Temperature ^oC	21	20	22
pH	7.79	7.68	7.99
Turbidity (ntu)	0.73	0.35	1.50
UV254 (cm^{-1})	0.029	0.025	0.037
DOC (mg/L)	2.34	2.11	2.56
SUVA	1.2	--	--
Total alkalinity (mg/L)	133	131	134
Total hardness (mg/L)	288	260	310
Bromide (mg/L)	0.08	0.05	0.10
Arsenic (mg/L)	0.003	0.002	0.004
Perchlorate (μg/L)	10	43	14
Sulfate (mg/L)	246	230	250
Chloride (mg/L)	84	75	94
TDS (mg/L)	619	602	642

This source water contained high concentrations of TDS, and correspondingly high concentrations of chloride and sulfate. Therefore, the presence of inorganic constituents that might compete with DOC for uptake on the MIEX® resin was extremely high, particularly with high levels of sulfate being present. DOC concentrations were relatively low, but bromide concentrations were significant in terms of capability for affecting the formation of disinfection by-products. There is not significant capability for removal of DOC by coagulation at this location due to (1) constraints imposed by the direct filtration process on applied coagulant doses, and (2) the moderately high alkalinity, which negatively impacts the ability to readily reduce pH to achieve more effective removal of DOC by coagulation. Therefore, a process such as MIEX® could potentially achieve greater DOC removal than is presently possible.

Bench-scale Testing

Bench scale dose-response testing was conducted to evaluate the uptake of DOC, UV254, arsenic, and perchlorate on the MIEX resin®. Results of these evaluations are shown on Figures 5.35 through 5.38. Arsenic and perchlorate were spiked to the raw water sample at a concentration of 20 μg/L. The resulting concentrations after the spike were 24 μg/L arsenic and 26 μg/L perchlorate. Dose-response analyses were also performed for bromide, but no removal occurred. This is in contrast with the results at HWTP and may reflect the much higher level of anion competition at AMSWTF, particularly in the presence of a high concentration of sulfate, which is a divalent anion.

Removal/reduction percentages for DOC and UV254 absorbance were much lower than at the other test sites (less than 60 and 30 percent, respectively), which may have been attributable to competing anions as was the case with bromide. The UV254 absorbance of the raw water at AMSWTF is much lower than for the other waters evaluated, which is reflected in the extremely low SUVA level. These differences in the characteristics of NOM may have also been factors that affected the efficiency of MIEX® treatment.

Reductions in UV254 appeared to reach a plateau. The limiting level of removal was less clear in the case of DOC, although the rate of incremental increase in removal with dose was small. No significant effect of increasing detention time was observed for DOC removal within a range of typical MIEX® doses.

Results of the arsenic analyses indicated increasing removals as MIEX® dose and contact time are increased. No maximum limit for arsenic removal was apparent for these tests, and thus it is possible that use of higher resin concentrations or increased regeneration frequency could have achieved higher removals. Perchlorate was more effectively removed under the test conditions that were applied, and appeared to reach a maximum removal rate of about 80 percent. Removals of both of these anions may have been affected by the high level of sulfate that was present and it is possible that different removals could be attained at other locations.

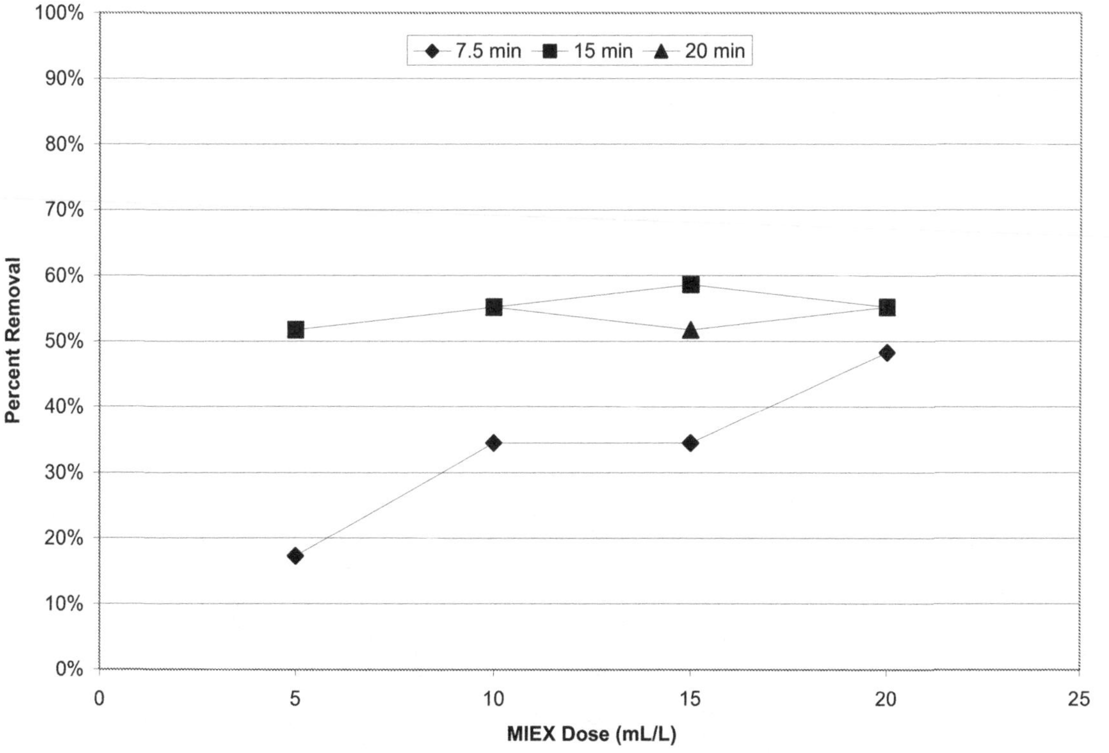

Figure 5.35 Percent UV254 absorbance reduction during dose-response testing at AMSWTF

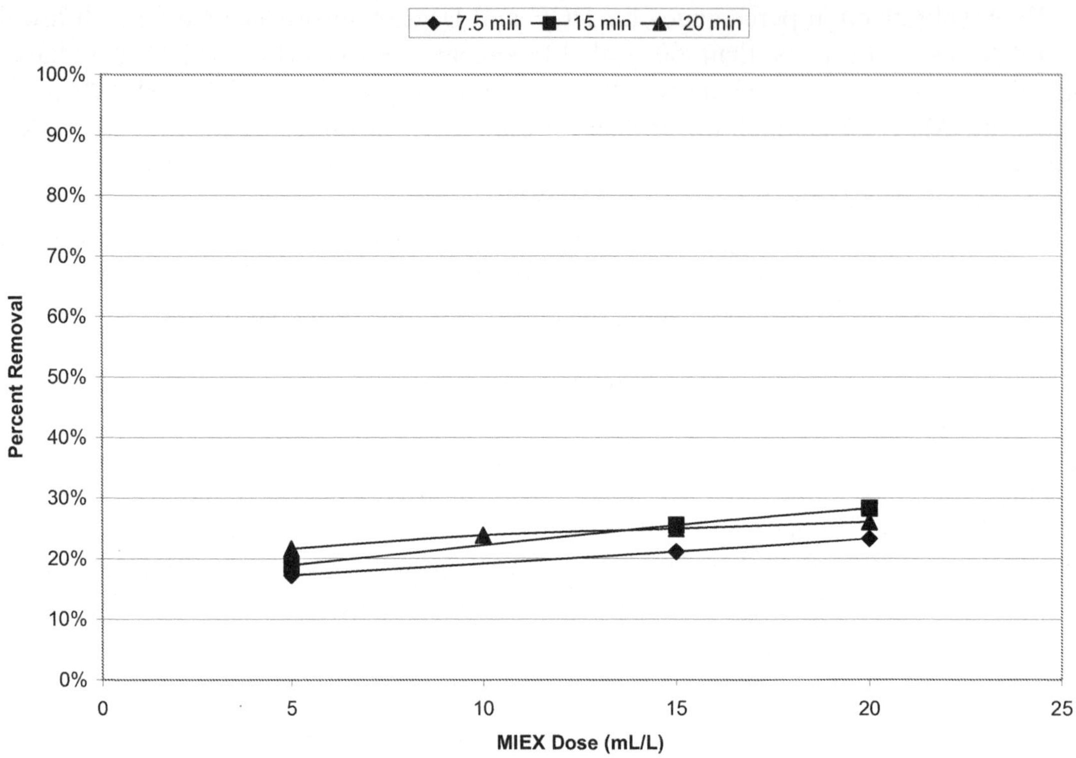

Figure 5.36 Percent DOC removal during dose-response testing at AMSWTF

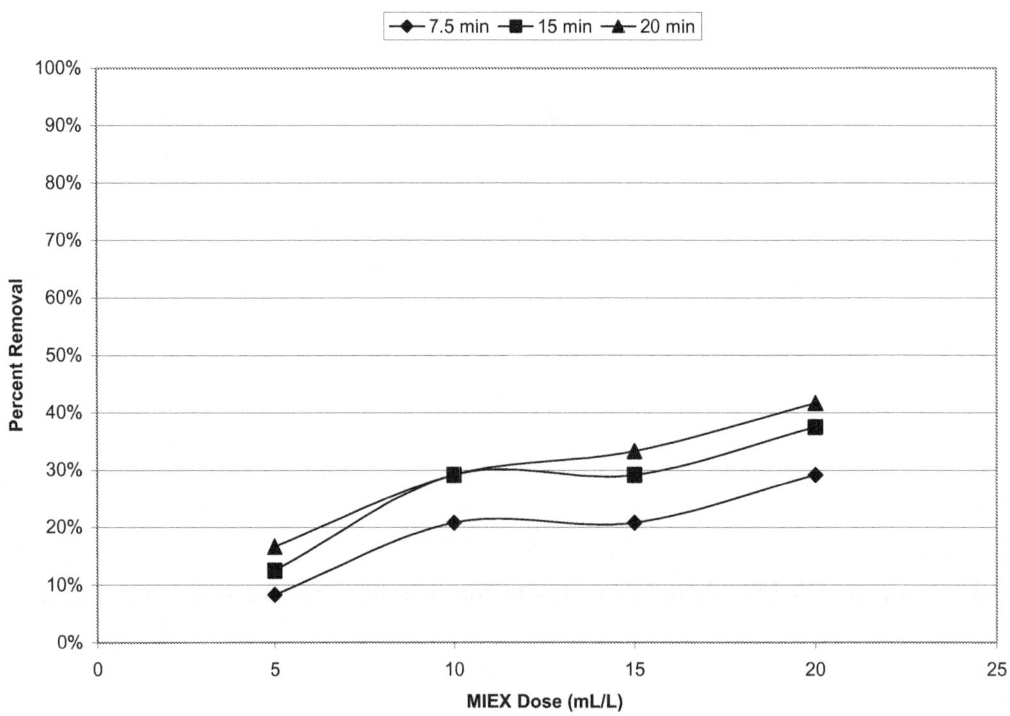

Figure 5.37 Percent arsenic removal during dose-response testing at AMSWTF

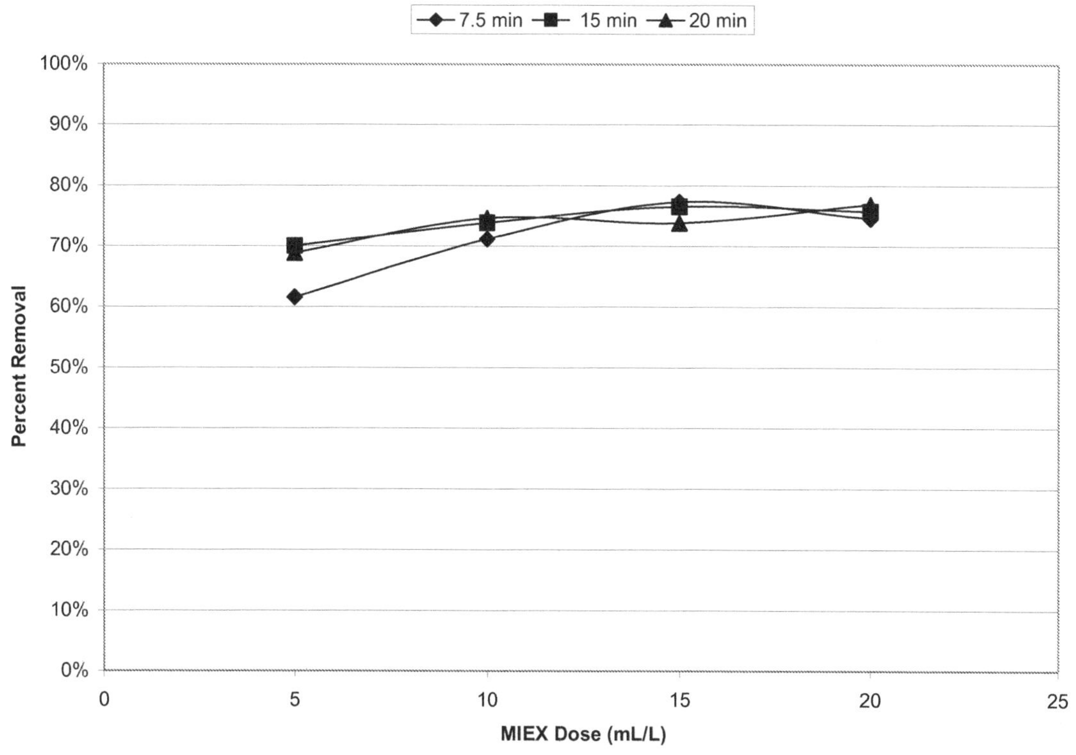

Figure 5.38 Percent perchlorate removal during dose-response testing at AMSWTF

Pilot Testing

Based on the results of the dose-response testing, pilot testing was performed at a resin concentration of approximately 20 mL/L, regeneration rates up to 10 percent, and a contact time of 15 minutes. A low bromide content salt was obtained for brine make-up in the pilot testing at this site. Specific operating conditions during the four consecutive weeks of pilot testing at AMSWTF are summarized below:

- Week 1 - 20 mL/L resin, 15 minute contact time, and 5 percent of flow sent to regeneration.
- Week 2 - 20 mL/L resin, 15 minute contact time, and 5 percent of flow sent to regeneration.
- Week 3 - 20 mL/L resin, 15 minute contact time, and 10 percent of flow sent to regeneration.
- Week 4 - 20 mL/L resin, 15 minute contact time, and 10 percent of flow sent to regeneration.

Results of UV254 absorbance and DOC reduction/removal testing are shown on Figures 5.39 through 5.42. These results reflect lower levels of removal than were achieved at the other test sites, and are consistent with the bench scale test results in this respect. UV254 and DOC reductions generally ranged from 40 to 70 percent and 20 to 45 percent, respectively. No discernible trend was observed in terms of changes in percent removal for different doses or percent regeneration conditions. This suggests that there may be little opportunity for additional removal of NOM once a certain level is achieved.

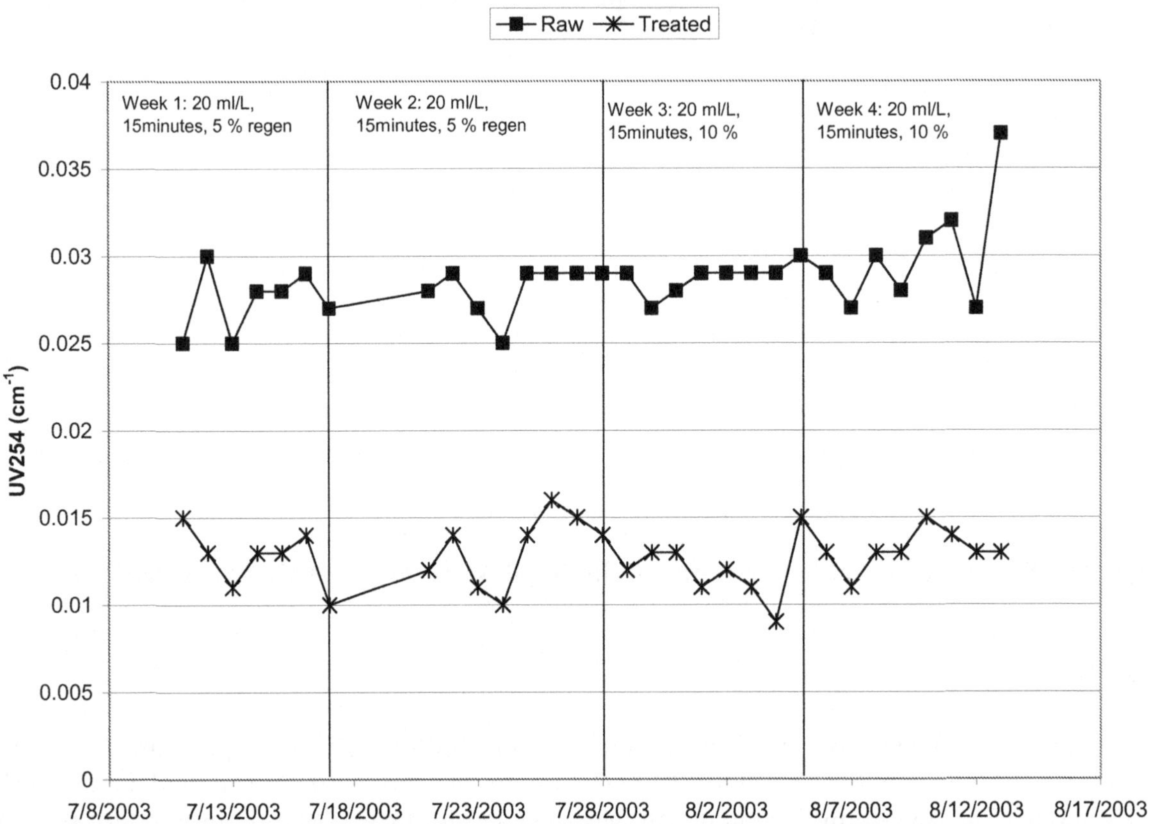

Figure 5.39 UV254 absorbance reduction during pilot plant operation at AMSWTF

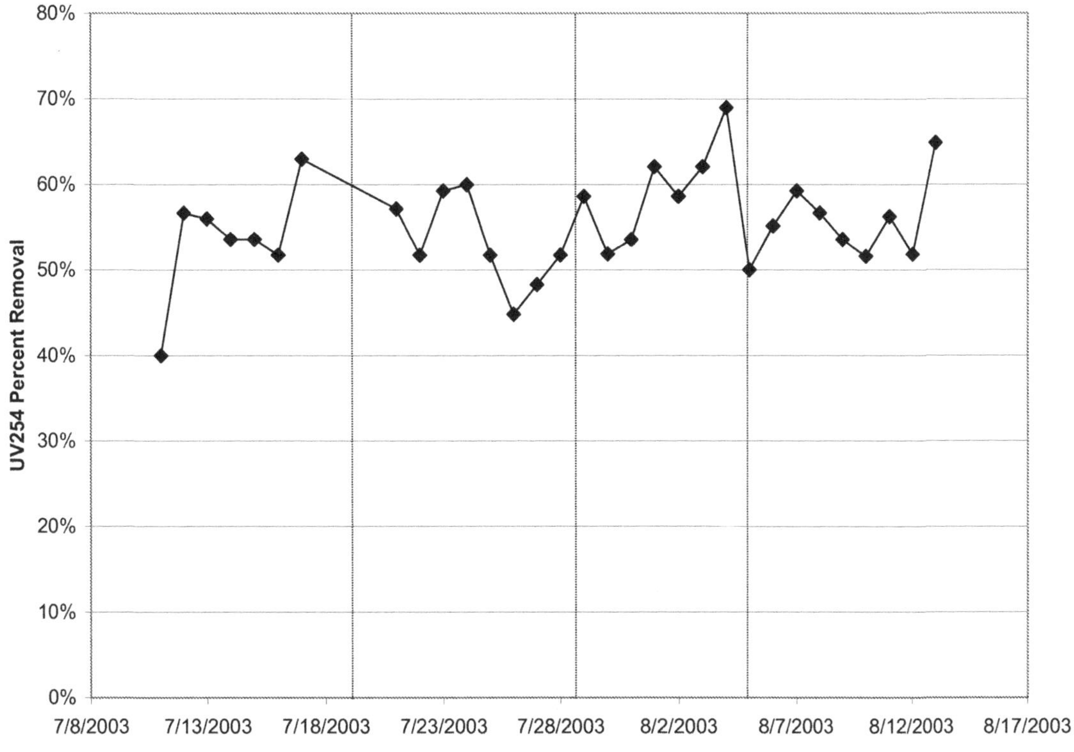

Figure 5.40 Percent UV254 absorbance reduction during pilot plant operation at AMSWTF

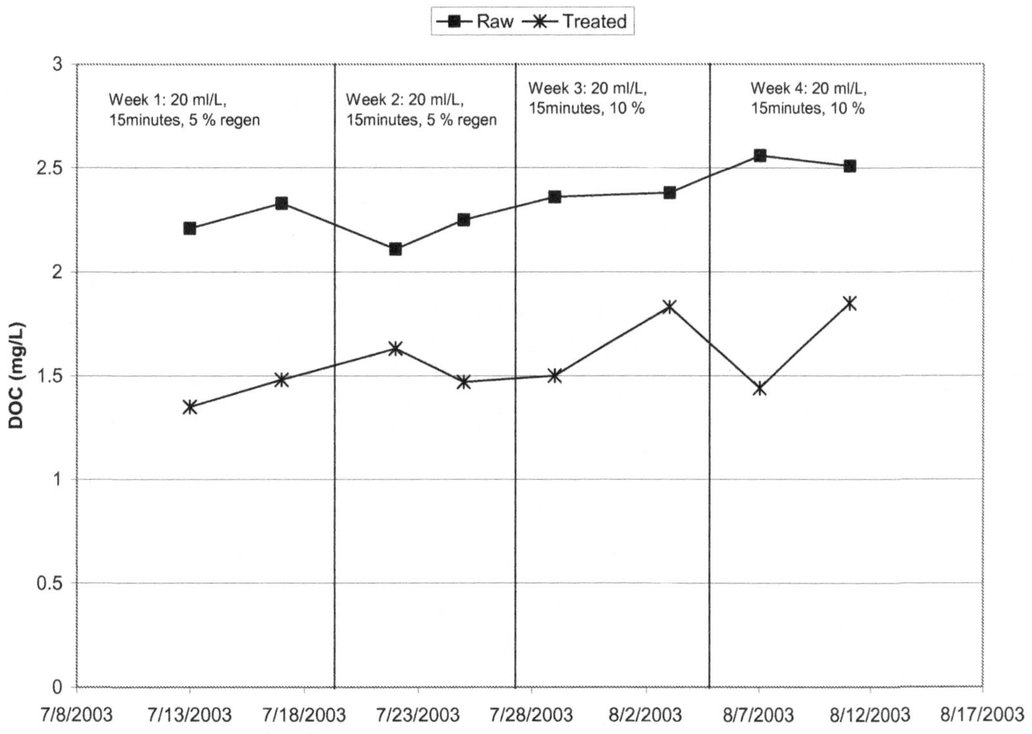

Figure 5.41 DOC removal during pilot plant operation at AMSWTF

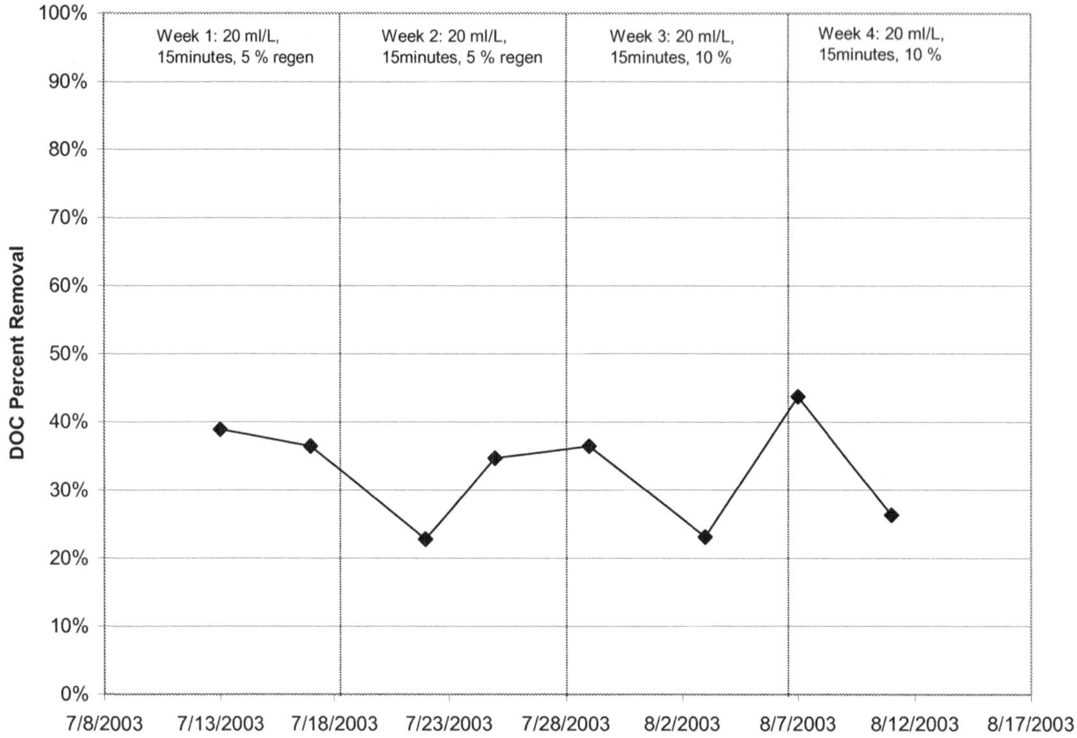

Figure 5.42 Percent DOC removal during pilot plant operation at AMSWTF

Bromide removal was not observed, and there were times during week 1 of testing when the bromide level in the raw water was lower than the treated bromide level, as shown on Figure 5.43. As the regenerant brine was prepared with low bromide content salt at this site, it is likely that the poor bromide removal results reflect the presence of significant anion competition for ion exchange sites rather than competition by bromide in the regenerant salt source.

Testing was also conducted to monitor the effect of the MIEX® process on arsenic and perchlorate. This testing indicated that arsenic was present in the raw water at relatively low levels of less than 4 μg/L and was not removed by the MIEX® process to a significant degree. The bench scale results suggested that some removal of arsenic was possible at higher levels of occurrence. Perchlorate removal was observed as shown on Figure 5.44. The percent removal was also not as high as was achieved in the bench scale tests, which may reflect the presence of lower concentrations of perchlorate in the feed water to the pilot plant.

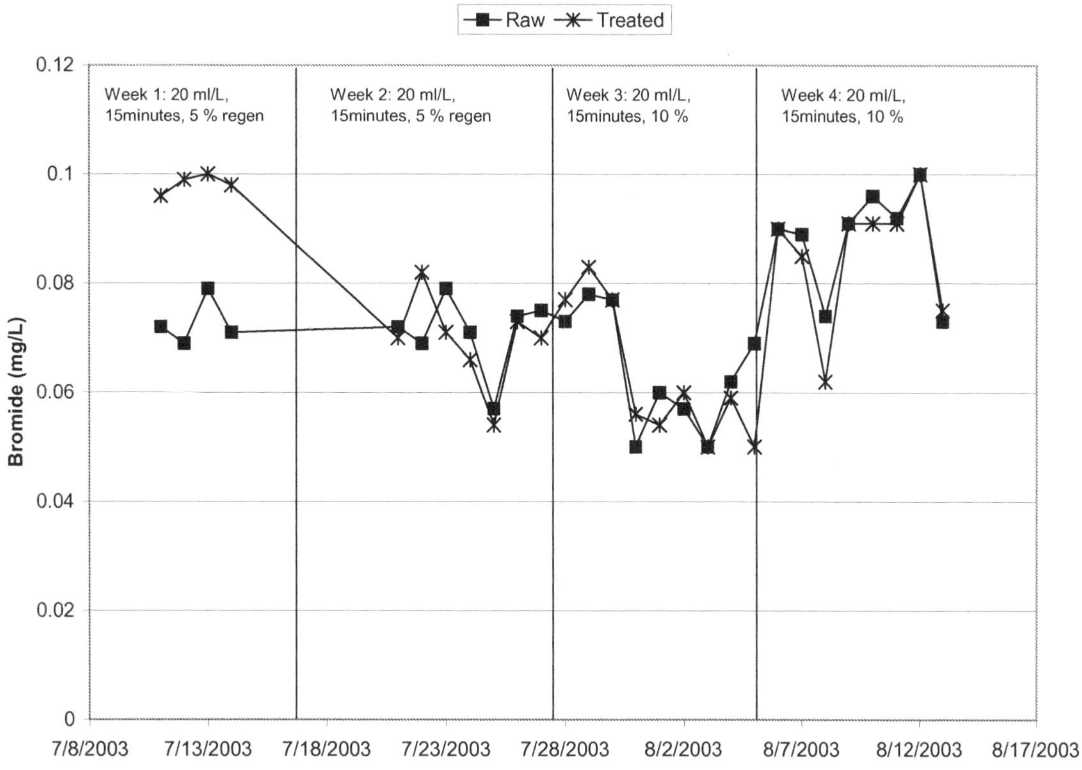

Figure 5.43 Bromide concentrations during pilot plant operation at AMSWTF

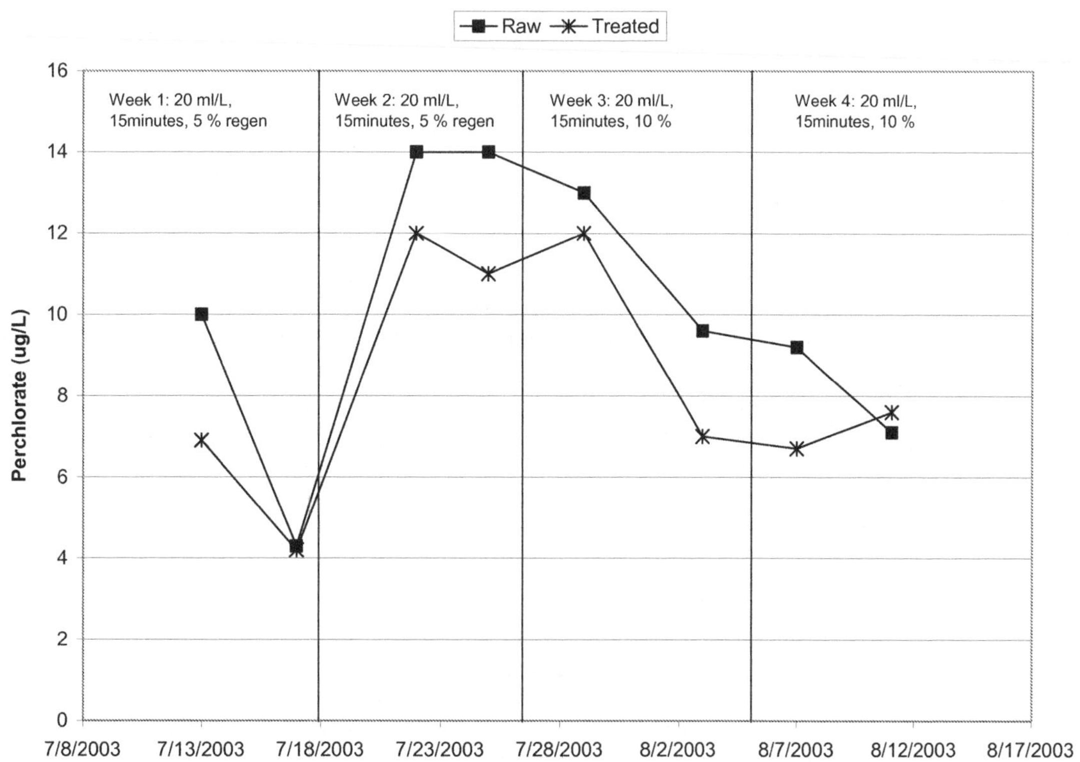

Figure 5.44 Perchlorate removal during pilot plant operation at AMSWTF

Detention Time Study

Detention time studies were performed each week to determine the effect of reducing the contact time to one half of the normal detention time. This change was implemented by bypassing one of the two contactors to reduce the contacting volume by half. A period of two hours was allowed for the change to take place across the pilot plant. The results of these studies are shown in Table 5.35 and are compared with the percent removals achieved at the contact time selected for the respective week of operation.

Table 5.35
Detention time study results at AMSWTF

	Week 1 - 20 mL/L 5% regeneration		Week 2 - 15 mL/L 5% regeneration		Week 3 - 15 mL/L 10% regeneration		Week 4 - 15 mL/L 10% regeneration	
	Percent removal 7.5 minutes	Percent removal 15 minutes	Percent removal 7.5 minutes	Percent removal 15 minutes	Percent removal 7.5 minutes	Percent removal 15 minutes	Percent removal 7.5 minutes	Percent removal 15 minutes
UV254 (cm^{-1})	44	54	48	53	38	60	50	56
DOC (mg/L)	32	36	29	35	31	23	25	26

Effect of the MIEX® Process on General Water Quality Characteristics

The effects of MIEX® treatment on general water quality characteristics are shown in Table 5.36.

Table 5.36
Average concentrations of water quality constituents at AMSWTF after MIEX® treatment

	pH	Total alkalinity (mg/L)	Total hardness (mg/L)	Sulfate (mg/L)	Chloride (mg/L)	TDS (mg/L)
Raw	7.79	133	288	246	84	619
Week 1	7.95	133	303	223	96	627
Week 2	7.83	131	287	233	113	631
Week 3	7.86	132	297	230	123	643
Week 4	7.80	132	270	215	175	677

A slight reduction in the sulfate concentrations was observed as expected. Chloride concentrations increased as a result of exchange for anions taken up by the resin; a significant increase occurred during Week 4, accompanied by a significant increase in TDS that may have been an indication of salt carry-through from the brine during regeneration. Hardness, alkalinity, and pH showed little change.

Figures 5.45 through 5.47 show the trends for sulfate, chloride, and TDS for the four weeks of pilot testing. The increase in chloride concentration across the process was generally in the range of 20 to 35 mg/L, with the exception of Week 4.

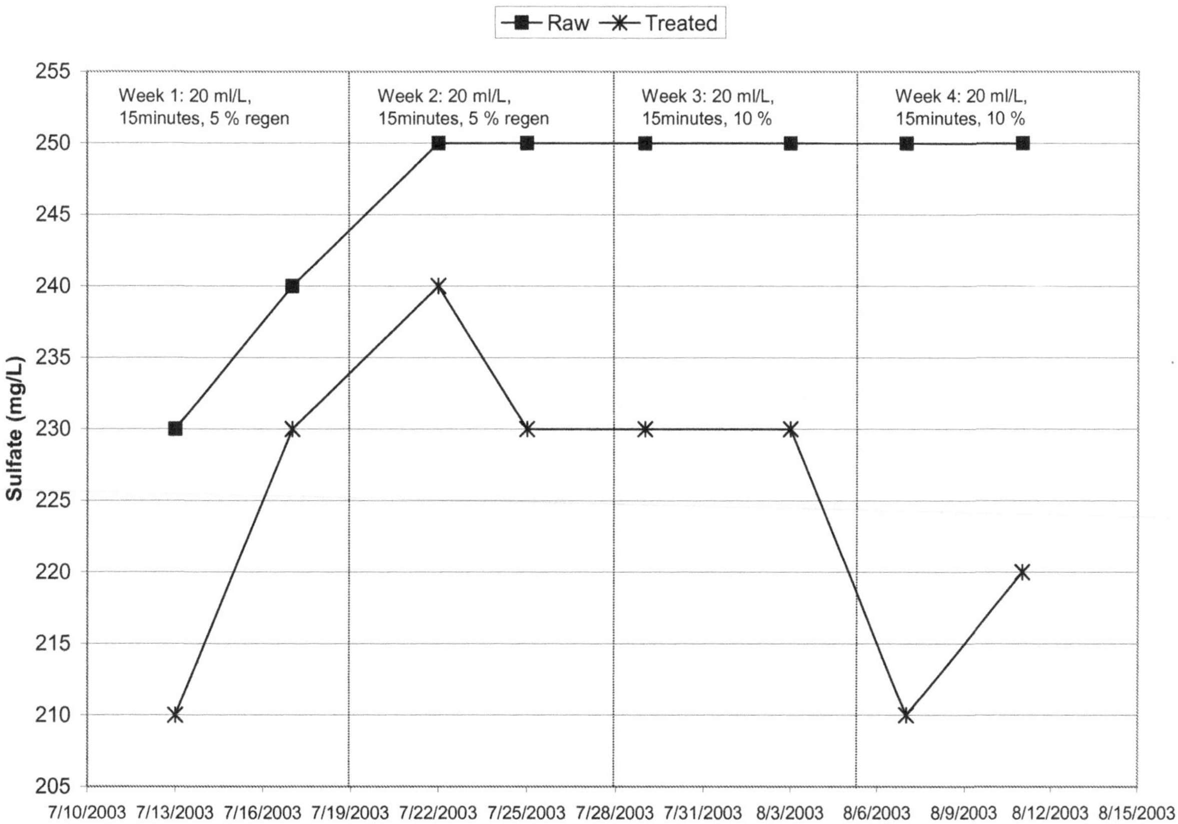

Figure 5.45 Sulfate concentrations during pilot plant operation at AMSWTF

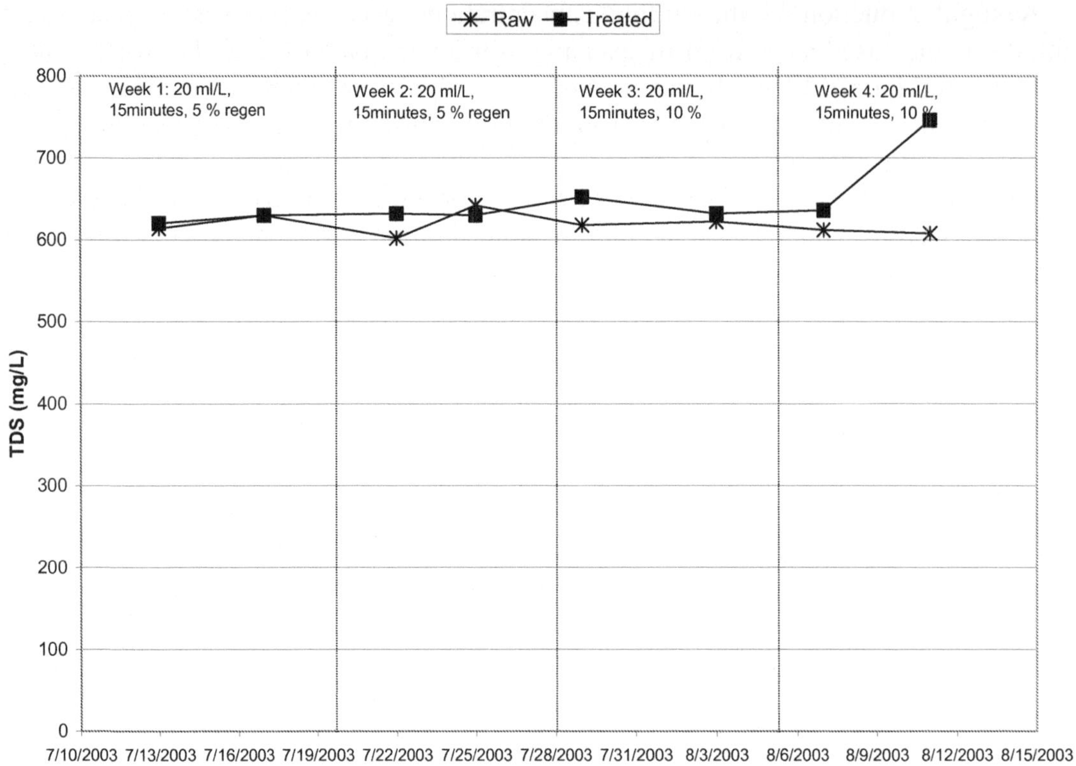

Figure 5.46 TDS concentrations during pilot plant operation at AMSWTF

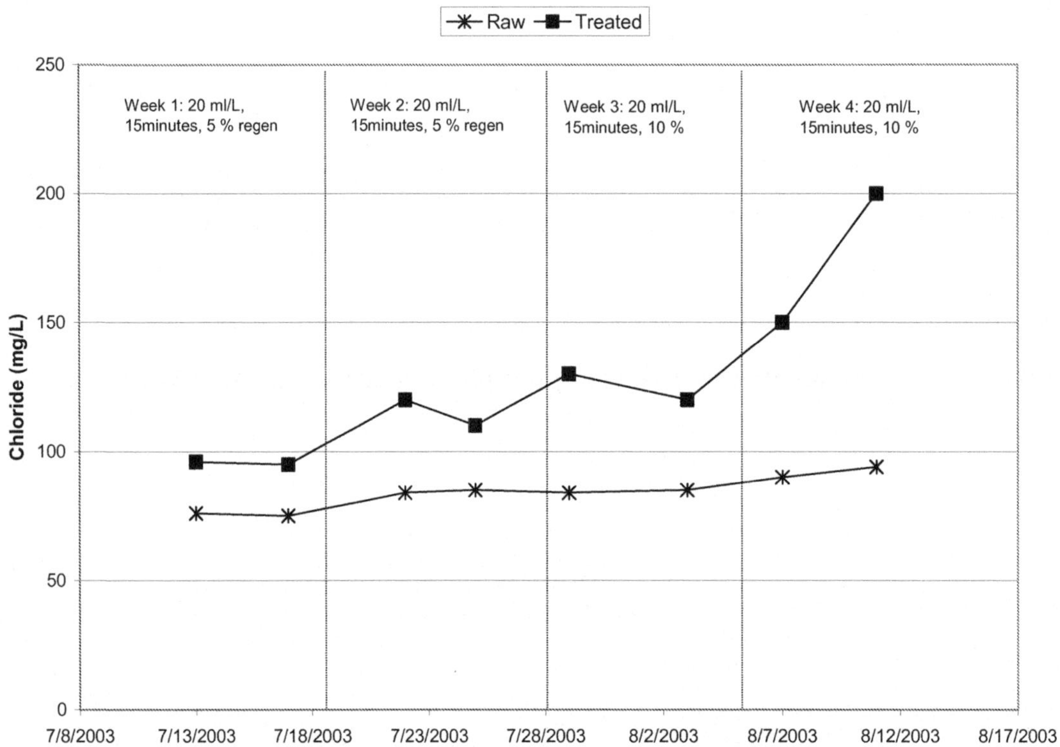

Figure 5.47 Chloride concentrations during pilot plant operation at AMSWTF

Evaluations of the Effect of MIEX® Pretreatment on Ozone Demand and Bromate Formation

Ozone decay studies were performed on raw water and MIEX® treated water during the Week 4 of pilot plant operation. Effluent from the MIEX® pilot plant was pumped to a 6 gpm ozone pilot plant to simulate a MIEX®–ozonation process sequence. Dissolved ozone residual profiles were collected from the top, middle, and bottom of contactor cells 1 through 5 to obtain sufficient data to fully develop the ozone decay profile. Ozone decay profiles were generated using several transferred ozone doses. Dissolved ozone residual measurements were taken using Standard Methods 4500-O_3 with potassium indigo trisulfonate.

A range of transferred ozone doses was used to generate a family of decay curves as shown in Figures 5.48 through 5.52. A resin concentration of 20 mL/L, 10 percent regeneration, and a 15 minute contact time were used in the MIEX® pilot plant for transferred ozone doses of 1.8 mg/L, 2.1 mg/L, 2.6 mg/L, and 3.1 mg/L, and a resin concentration of 10 mL/L, 10 percent regeneration, and a 15 minute contact time were used for 1.5 mg/L and 1.8 mg/L transferred ozone doses.

A reduction in DOC that would affect the ozone decay relationship was the anticipated result following MIEX® pretreatment. The effects on initial ozone demand and half life are summarized in Table 5.37. While the initial ozone demand was typically greater in the MIEX® treated water, suggesting an increase in rapidly oxidized material, the half life was longer, suggesting the removal of materials that impose a subsequent demand. One possible explanation for the higher initial demand was carryover of resin into the ozone contactor. The increase in half life can likely be attributed to DOC reduction from MIEX® pretreatment, and could allow target CT values to be achieved at lower ozone doses.

The results shown on Figures 5.48 through 5.52 reflect the slightly higher initial demand for ozone followed by a period of slower decay. In terms of exposure to ozone for disinfection, the overall effect of MIEX® pretreatment was beneficial. The effect of a competing initial demand would be less clear for achieving oxidation goals with ozone. The incorporation of a clarification step between the MIEX® and ozonation processes, as might be possible at some plants, would eliminate the carry-through of resin from the MIEX® process, thereby limiting its impact on initial ozone demand.

Table 5.37
Ozone demand and half-life results for raw water and MIEX® pretreatment studies

Transferred ozone dose (mg/L)		Temperature (°C)		Initial Ozone demand (mg/L)		Half life (minute)	
Raw water	MIEX® treated water	Raw water	MIEX® treated water	Raw water	MIEX® treated water	Raw water	MIEX® treated water
1.45	1.46	23.0	22.4	0.52	0.70	2.11	5.35
1.88	1.79	21.6	22.4	0.82	0.92	4.38	7.79
1.88	1.83*	21.6	22.5*	0.82	0.89*	5.43	8.47
2.07	2.18	21.0	21.8	0.93	1.05	5.43	9.79
2.53	2.64	21.3	22.4	1.22	1.11	6.79	11.66
3.13	3.17	21.1	22.4	1.31	1.35	8.15	10.38

*MIEX® condition of 10 mL/L, 10 percent regeneration, 15 minutes contact time

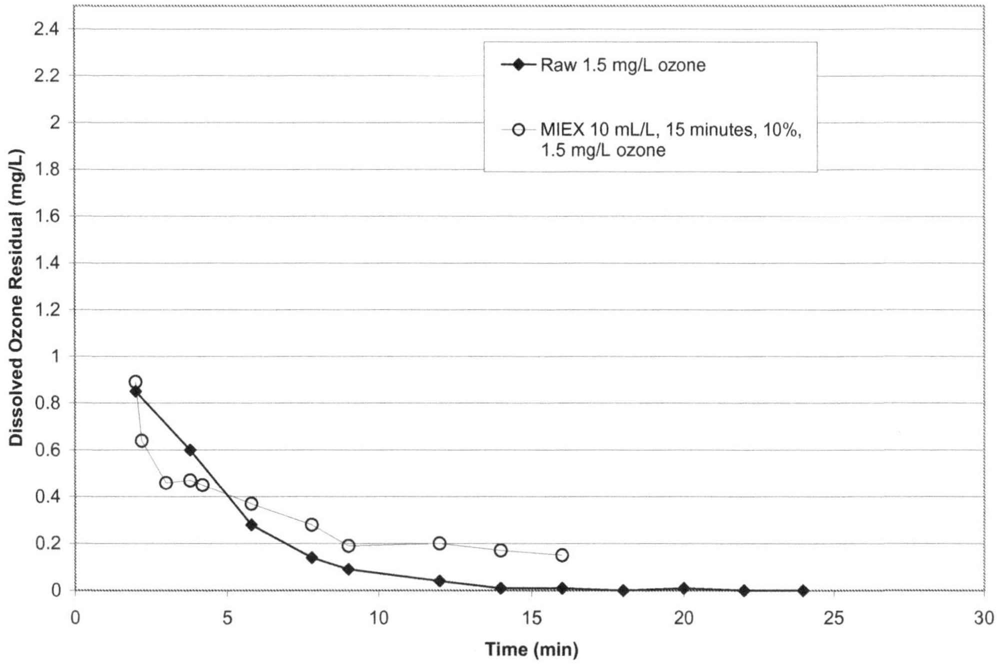

Figure 5.48 Dissolved ozone residuals in raw and MIEX® treated water (1.5 mg/L ozone, 20 mL/L resin , 15 minutes contact time, 10 % regeneration)

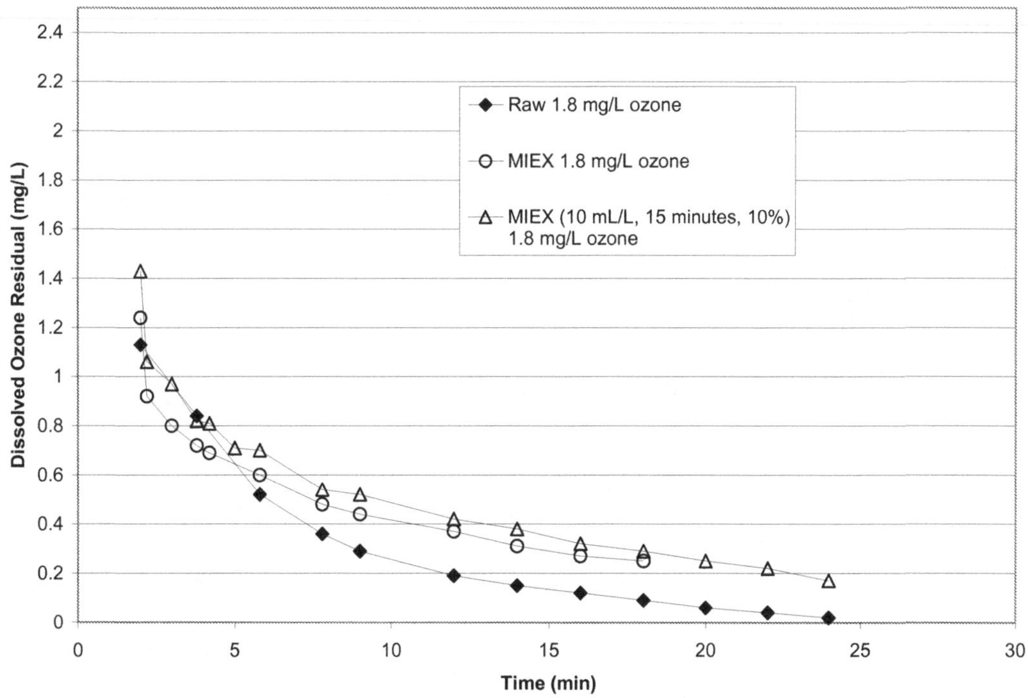

Figure 5.49 Dissolved ozone residuals in raw and MIEX® treated water (1.8 mg/L ozone, 20 mL/L resin, 15 minutes contact time, 10 % regeneration)

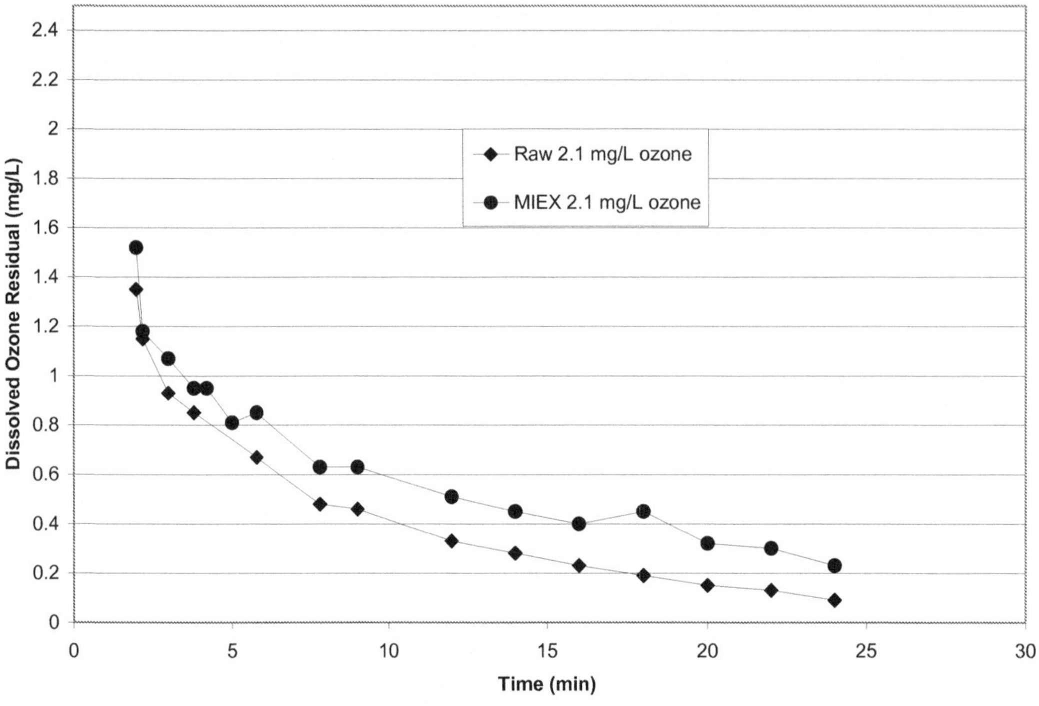

Figure 5.50 Dissolved ozone residuals in raw and MIEX® treated water (2.1 mg/L ozone, 20 mL/L resin, 15 minutes contact time, 10 % regeneration)

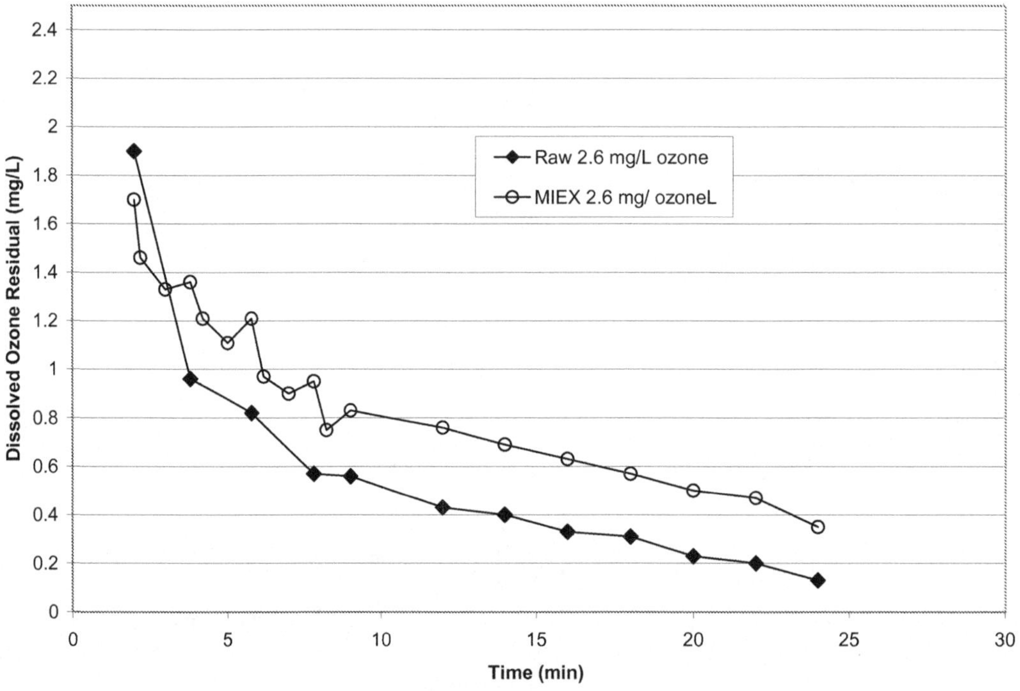

Figure 5.51 Dissolved ozone residuals of raw and MIEX® treated water (2.6 mg/L ozone, 20 mL/L resin, 15 minutes contact time, 10 % regeneration)

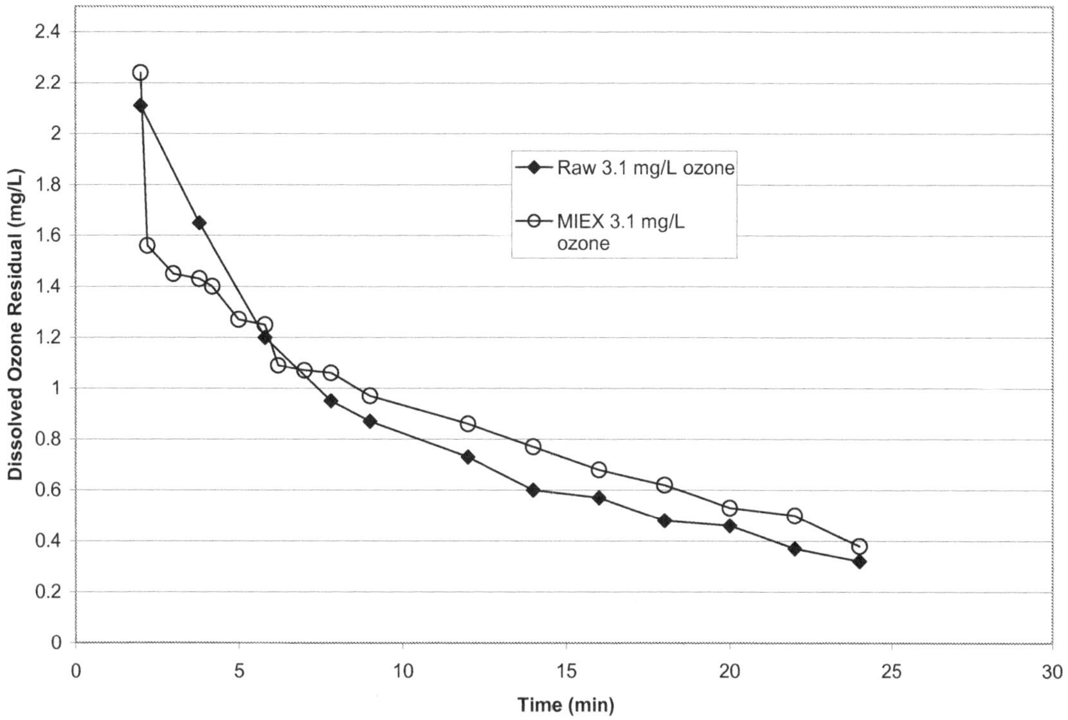

Figure 5.52 Dissolved ozone residuals in raw and MIEX® treated water (3.1 mg/L ozone, 20 mL/L resin, 15 minutes contact time, 10 % regeneration)

While the results of previous tests indicate that the MIEX® process was not effective in removing bromide from the AMSWTF water source, some bromate values following ozone treatment were unusually low (at or near the detection limit), and therefore further investigation may be warranted. Table 5.38 summarizes bromide and bromate concentrations following the application of ozone. The samples to be analyzed for bromate were collected from the effluent of the ozone contactor, and were not quenched, which resulted in higher bromate formation as a result of slightly longer exposure to ozone.

Table 5.38
Bromide and bromate results for ozone only and for MIEX® followed by ozone (20 mL/L MIEX®, 15 minute contact time, 10 percent regeneration)

Transferred ozone dose (mg/L)		Ozone only		MIEX® and ozone		
Raw water	MIEX® treated water	Raw water bromide (mg/L)	Post-ozone bromate (mg/L)	Raw water bromide (mg/L)	Post MIEX® bromide (mg/L)	Post MIEX®, post-ozone Bromate (mg/L)
1.45	1.46	0.067	0.010	0.089	0.085	0.0099
1.88	1.79	0.062	0.017	0.089	0.085	0.018
1.88	1.83*	0.062	0.010	0.092	0.091	0.005**
2.07	2.18	0.05	0.010	0.074	0.062	0.028
2.53	2.64	0.05	0.010	0.074	0.062	0.058
3.13	3.17	0.067	0.069	0.092	0.091	0.005**

*MIEX® condition of 10 mL/L, 10 percent regeneration, 15 minutes contact time
** Detection limit of method

Filtration Studies

Pilot-scale filtration test results conducted using MIEX® pilot plant effluent showed a much higher rate of head loss accumulation than for the full-scale plant, with a value of 5.5 feet observed during one filter run after only 16 hours of operation. The raw water supply for the AMSWTF has extremely low turbidity, and filter runs are typically much longer than the relatively short runs experienced during the pilot testing. It is likely that the main factor contributing to the higher head loss following the MIEX® process is the increased particulate content that results from the carryover of some of the smaller MIEX® resin particles in the process effluent. This would be unlikely to occur at a plant that employs an intermediate clarification step prior to filtration, since most of these particles will be removed prior to filtration.

Analyses of THM and HAA5 Formation

The effects of NOM removal by the MIEX® process on the formation of THMs and HAA5 were evaluated during two time periods. Disinfection by-product formation was assessed under UFC, with samples incubated at 20°C and pH 8.0. Pilot plant effluent was used as the source of treated water, and raw water was obtained at the pilot plant influent. A contact period of 15 minutes, a regeneration rate of 5 percent, and a resin concentration of 20 mL/L, were used for the first test period. For the second test period, the regeneration rate was increased to 10 percent. Key testing conditions for the test aliquots during the two test periods are summarized in Tables 5.39 and 5.40, and the resulting THM and HAA5 formations are shown in Tables 5.41 through 5.44.

Table 5.39
Treatment conditions at AMWTP during Week 1

	MIEX® concentration (mL/L)	Contact time (minutes)	MIEX® regeneration rate (percent)	pH	DOC (mg/L)
Raw	-	-	-	8.03	2.6
MIEX®	20	15	5	8.03	2.3

Table 5.40
Treatment conditions at AMSWTF during Week 4

	MIEX® concentration (mL/L)	Contact time (minutes)	MIEX® regeneration rate (percent)	pH	DOC (mg/L)
Raw	-	-	-	7.73	2.5
MIEX®	20	15	10	7.74	1.7

Table 5.41
THM concentrations from DBP formation potential experiments at AMSWTF – Week 1

	Chlorine residual (mg/L)			Cl_3CH (μg/L)			$BrCl_2CH$ (μg/L)			Br_2ClCH (μg/L)			Br_3CH (μg/L)			Total THM (μg/L)		
	2 hour	1 day	3 day	2 hour	1 day	3 day	2 hour	1 day	3 day	2 hour	1 day	3 day	2 hour	1 day	3 day	2 hour	1 day	3 day
Raw water	2.4	1.7	1.0	12.3	42.6	67.4	14.5	28.5	35.4	13.9	21.9	23.3	1.2	1.7	1.9	41.9	94.7	128
Post MIEX®	1.8	1.3	0.7	4.4	20.0	37.1	7.2	18.5	26.5	11. 6	20.7	23.8	2.2	3.5	4.1	25.4	62.7	91.5

Table 5.42
THM concentrations from DBP formation potential experiments at AMSWTF – Week 4

	Chlorine residual (mg/L)			Cl_3CH (µg/L)			$BrCl_2CH$ (µg/L)			Br_2ClCH (µg/L)			Br_3CH (µg/L)			Total THM (µg/L)		
	2 hour	1 day	3 day	2 hour	1 day	3 day	2 hour	1 day	3 day	2 hour	1 day	3 day	2 hour	1 day	3 day	2 hour	1 day	3 day
Raw water	2.5	1.8	1.2	11.7	42.0	67.4	14.4	28.4	35.5	13.7	22.4	23.6	1.4	1.9	2.2	41.2	94.7	128.7
Post MIEX®	2.4	1.6	1.3	3.9	19.6	35.8	6.2	16.1	23.5	10.0	17.8	20.9	2.1	3.2	3.8	22.2	56.7	84

Table 5.43
HAA5 concentrations from DBP formation potential experiments at AMSWTF – Week 1

	ClAA (µg/L)			BrAA (µg/L)			Cl_2AA (µg/L)			Br_2AA (µg/L)			Cl_3AA (µg/L)			Total HAA5 (µg/L)		
	2 hour	1 day	3 day	2 hour	1 day	3 day	2 hour	1 day	3 day	2 hour	1 day	3 day	2 hour	1 day	3 day	2 hour	1 day	3 day
Raw water	<2	<2	<2	<2	<2	<2	3.7	11.4	17.1	<2	2.7	2.9	5.0	8.3	10.3	8.7	22.4	30.3
Post MIEX®	<2	<2	<2	<2	<2	<2	<2	4.7	7.6	<2	2.5	3.0	2.8	4.1	5.5	2.8	11.3	16.1

Table 5.44
HAA5 concentrations from DBP formation potential experiments at AMSWTF – Week 4

	ClAA (μg/L)			BrAA (μg/L)			Cl_2AA (μg/L)			Br_2AA (μg/L)			Cl_3AA (μg/L)			Total HAA5 (μg/L)		
	2 hour	1 day	3 day	2 hour	1 day	3 day	2 hour	1 day	3 day	2 hour	1 day	3 day	2 hour	1 day	3 day	2 hour	1 day	3 day
Raw water	<2	<2	<2	<2	<2	<2	3.6	9.6	15.8	<2	2.4	3.0	5.1	8.7	12.1	8.7	20.7	30.9
Post MIEX®	<2	<2	<2	<2	<2	<2	<2	3.9	7.5	<2	2.3	2.8	2.8	4.1	5.5	2.8	10.3	15.8

While DOC reductions by the MIEX® process at AMSWTF were modest, the reduction in THM formation was significant, and the MIEX® process could therefore contribute to compliance with future regulatory requirements. However, this process did not exhibit THM formation levels that were less than 0.080 mg/L in this work and may not be adequate for achieving the goals for THM at this location by itself. The availability of downstream ozonation, coagulation, and filtration steps could further enhance these removals, but further analysis would be required to confirm the effect of process combinations. Formation of HAA5 tended to be relatively low and appears to be less of a factor with regard to future regulatory compliance at this location.

Analyses of Resin Activity and Buildup of DOC in the Regenerant Brine

Testing was performed to compare the effectiveness of virgin resin, regenerated resin collected at startup of the pilot plant, and regenerated resin collected at the end of the four-week period of pilot plant operation. Parallel dose-response tests were used to assess the uptake of UV254 and DOC from a common raw water sample collected from the pilot plant. The concentration of DOC in this water was 2.2 mg/L, and the UV254 absorbance was 0.037 cm^{-1}. Results of the dose-response tests, which are summarized in Table 5.45, do not indicate a loss of activity for the resin during the test period. The 3 mL/L result after 4 weeks of operation is abnormally high and appears to be an outlier. Results for DOC removals at the other concentration give a more general indication of the activity and the trends for UV254 reduction also give indications of maintained capacity.

Table 5.45
Evaluation of resin activity at AMSWTF

Utility	MIEX® dose (mL/L)	UV254 absorbance reduction (%)	Percent DOC removal (%)
Virgin resin	1	16	7
	2	35	22
	3	41	44
Regenerated resin at the beginning of operation (one week of startup)	1	27	5
	2	35	20
	3	38	--
Regenerated resin after five weeks of operation (four weeks of operation)	1	11	8
	2	32	33
	3	49	73

Buildup of DOC as a consequence of re-use of the brine was assessed by analyzing the brine for DOC concentration at the end of each week before preparing a fresh regenerant batch. The results of these analyses are shown in Table 5.46. These levels are relatively low and reflect the low DOC content of the raw water, as was the case at MWTP and FTWTP.

Table 5.46
DOC buildup in the regenerant brine at AMSWTF

Week of pilot operation	DOC concentration at end of week of operation (g/L)
1	0.5
2	0.6
3	1.1
4	1.2

CHAPTER 6
RESULTS OF MOLECULAR WEIGHT FRACTIONATION STUDIES AND DISINFECTION BY-PRODUCT FORMATIONS

This chapter discusses the results of the disinfection by-product formation studies and the molecular weight and hydrophilic/hydrophobic fractionations. These studies were performed twice at each site. The disinfection by-product formation results are reported as four trihalomethane compounds and nine haloacetic acid compounds; whereas only the five regulated HAA5 compounds are reported in Chapter 5.

REMOVAL OF DOC AND UV-ABSORBING MATERIAL BY MIEX® RESIN

The concentrations of DOC and UV-absorbing materials were measured before and after MIEX® treatment to assess the ability of the resin to remove NOM. Figure 6.1 summarizes DOC removals at all four WTPs; the values shown are averages of Week 1 and Week 4 concentrations. DOC removals with the MIEX process were substantial at MWTP (63 percent), HWTP (64 percent), and FTWTP (57 percent), but relatively low at AMSWTF (22 percent). The lower removal at AMSWTF was expected because of the lower SUVA and the presence of high TDS concentrations in the raw water that compete for sites on the resin (see discussion of AMSWTF testing in Chapter 5).

Figure 6.2 shows the decrease in UV-absorbing materials by treatment based on averages of Week 1 and Week 4 values. The MIEX® process removed 75 percent of the UV-absorbing material at the FTWTP, 78 percent at the HWTP, and 74 percent at the MWTP. Results at the AMSWTF were lowest at 56 percent, which was most likely attributable to the higher concentration of TDS and lower SUVA in the raw water. The SUVA value at each utility decreased after MIEX® treatment, indicating that the MIEX process removes more of the UV-absorbing component of the DOC.

The raw water supplies for MWTP and AMSWTF are low in turbidity and therefore either do not require coagulation/sedimentation for turbidity removal, or require only a small amount of coagulant. Therefore, coagulation evaluations were not appropriate at these locations. The results for the other two locations show that MIEX® treatment achieved better DOC removal than coagulation at FTWTP, but that MIEX® treatment and coagulation produced similar results at HWTP. Maximum DOC removal was achieved by MIEX® treatment followed by coagulation at both locations, even when much lower dose of coagulant was applied for turbidity removal.

The MIEX® process removes as much or more of the UV-absorbing material from the raw water as coagulation. The greatest removal rate is achieved with MIEX® treatment followed by coagulation, even at a lower coagulant dose.

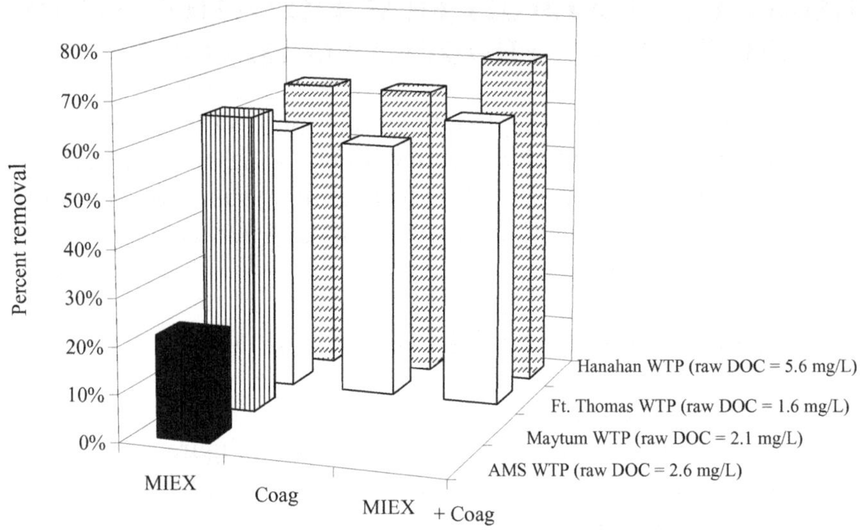

Figure 6.1 Average percent removal of DOC at each utility by treatment process

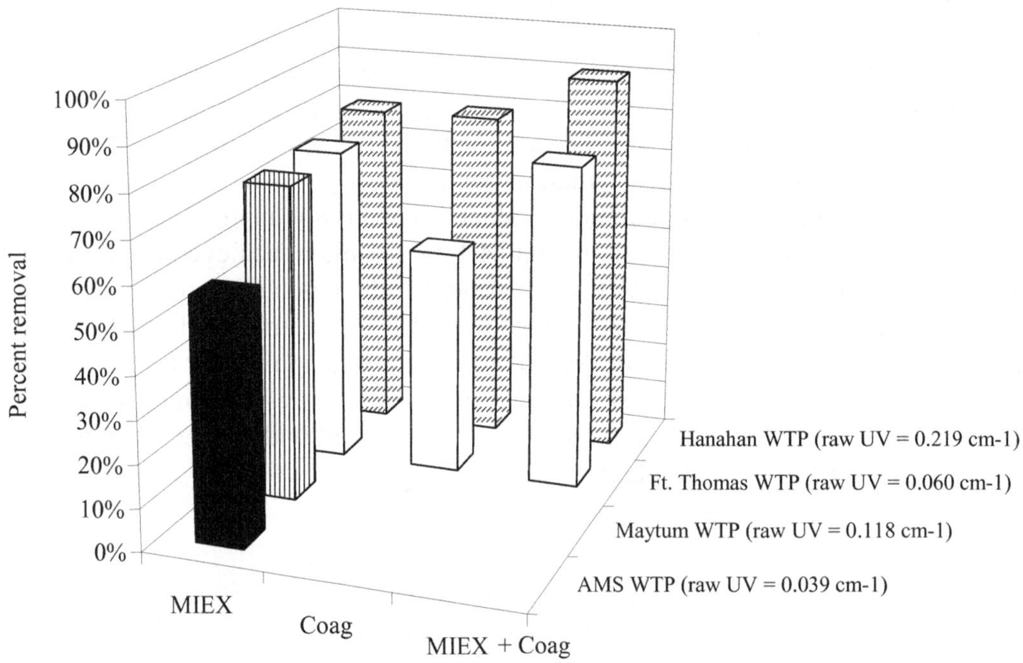

Figure 6.2 Average percent removal of UV-absorbing material at each site

XAD FRACTIONATION

The results of XAD fractionation at each utility are summarized on Figure 6.3. The hydrophobic acids are shown as HPOA, the hydrophilic acids as HPIA, and transphilic acids as TPHA. In presenting these results, it should be noted that DOC values less than 0.5 mg/L are less reliable due to the analytical limitations of the instrument. While general trends may be seen; values in this range should not be regarded as absolute.

The values shown on Figure 6.3 are averages of Week 1 and Week 4 concentrations for each utility. It was observed that HPOA comprises 64 percent of the organic matter in water treated at the MWTP, 54 percent at HWTP, 43 percent at FTWTP and 33 percent of the DOC at AMSWTF, an observation that is consistent with the trend of DOC removals at these locations. For example, water treated at the MWTP and HWTP contained the highest percentage of HPOA, consistent with the higher level of DOC removal at these locations. The results shown in this figure indicate that MIEX® treatment tended to remove a portion of each acid fraction.

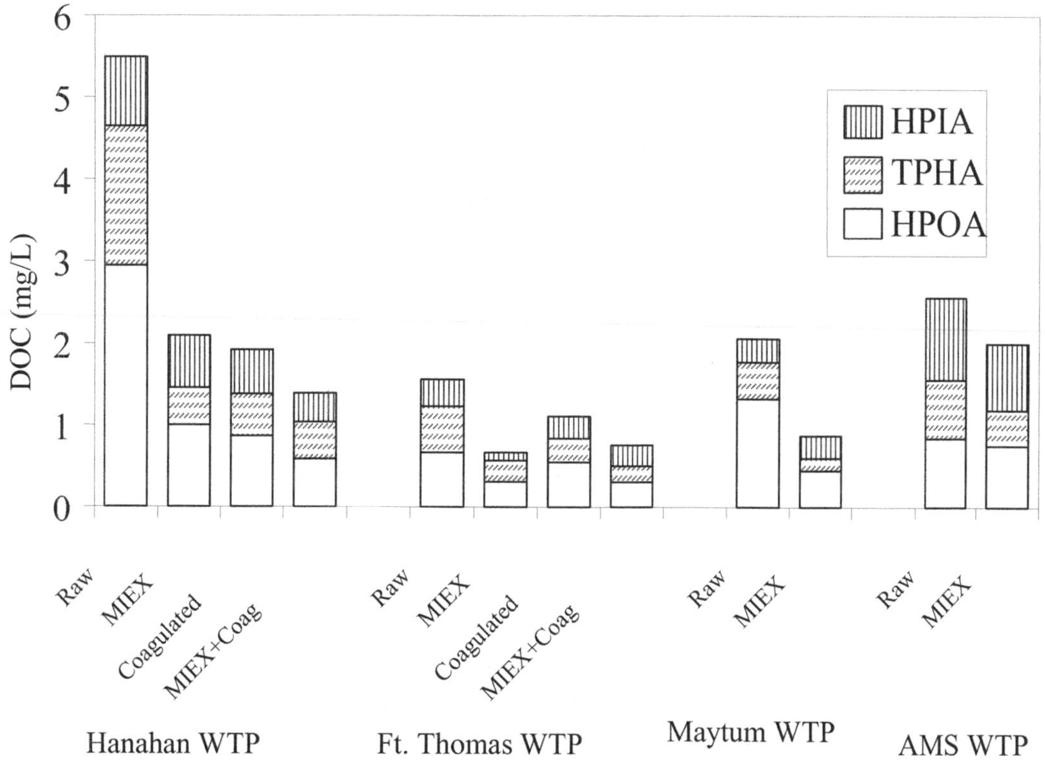

Figure 6.3 Summary of DOC concentrations for each XAD fraction

Figure 6.4 shows a comparison of removals of different organic acid fractions by MIEX® treatment, by coagulation, and by MIEX® treatment followed by coagulation at the HWTP, where coagulation was a part of the study. This site was selected for these comparisons because the higher DOC levels at this site allowed more accurate characterizations of the respective fractions. MIEX® treatment alone produced results that were generally similar to those for coagulation alone, but MIEX® treatment followed by coagulation achieved the highest overall removals of DOC as well as of each organic acid fraction of DOC, including much improved removal of the hydrophilic acids.

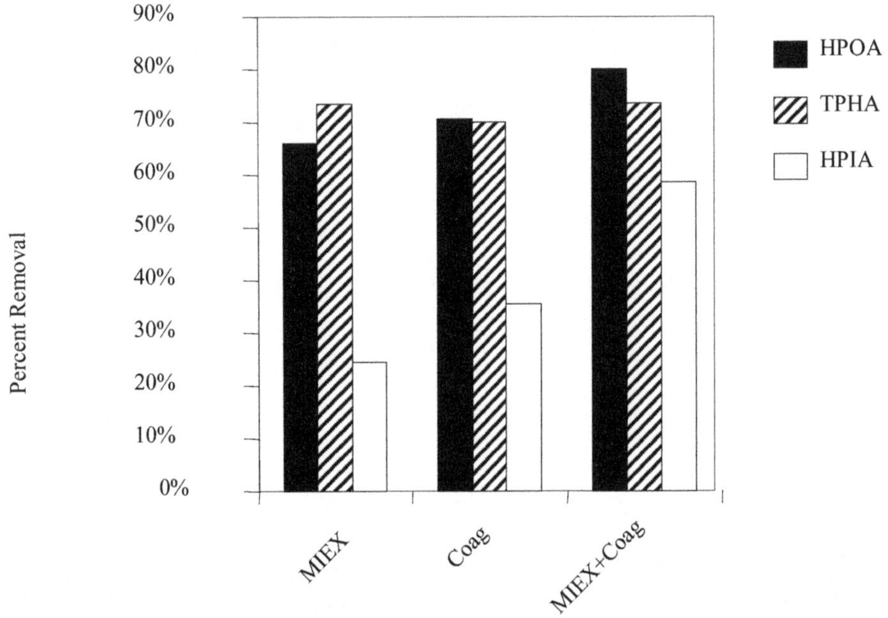

Figure 6.4 Effect of MIEX® and coagulation on the removal of organic acid fractions at HWTP

MOLECULAR WEIGHT FRACTIONATION

Figure 6.5 shows the results of the molecular weight fractionation at each utility. At the FTWTP, the MWTP, and the AMSWTF, the largest molecular weight fraction consisted of DOC from 1,000 to 10,000 NMW, with much less of the DOC in the molecular weight range greater than 10,000 NMW. In contrast, the 10,000 to 30,000 NMW fraction was the largest at the HWTP site. Analytical results for molecular weight fractions of greatest concentration are most meaningful. For this reason, greatest credibility is given to results in treated water samples for the HWTP.

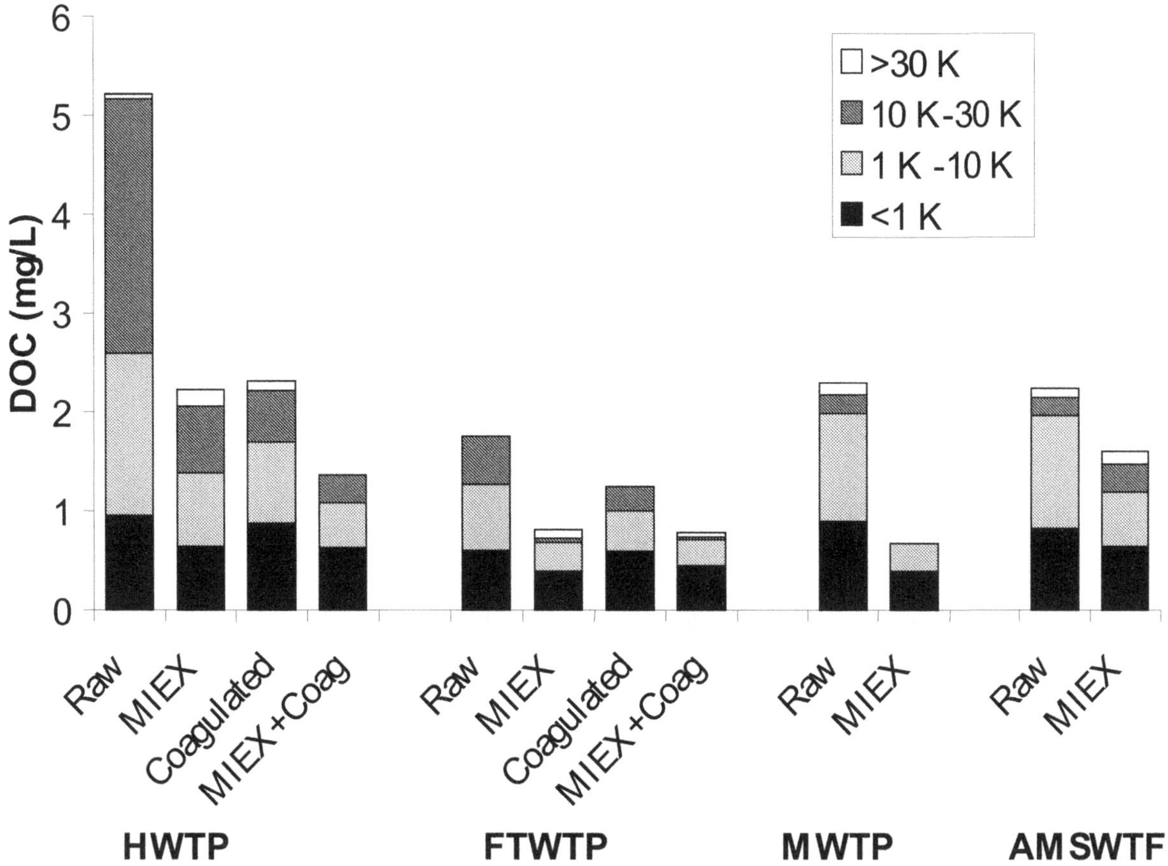

Figure 6.5 Summary of average DOC concentrations of each molecular weight fraction

Results from the study indicated that MIEX® treatment and coagulants remove different molecular weight fractions of NOM to varying degrees. Figure 6.6 shows a comparison of removals by MIEX® treatment, by coagulation, and by MIEX® followed by coagulation at the HWTP, which was selected for analysis because of the higher DOC concentrations in the respective fractions. The results show that MIEX® treatment and coagulation removed substantial amounts of the 1,000 through 10,000 and the 10,000 through 30,000 NMW fractions, but that MIEX® treatment also removed part of the fraction less than 1,000 NMW, which was not removed by coagulation. Therefore, MIEX® treatment and coagulation appear to remove slightly different NOM fractions. This was supported by differences in the extent to which THMs and HAAs were formed and by differences in UV254 reduction following treatment by the respective processes to comparable levels of DOC, as discussed in Chapter 5.

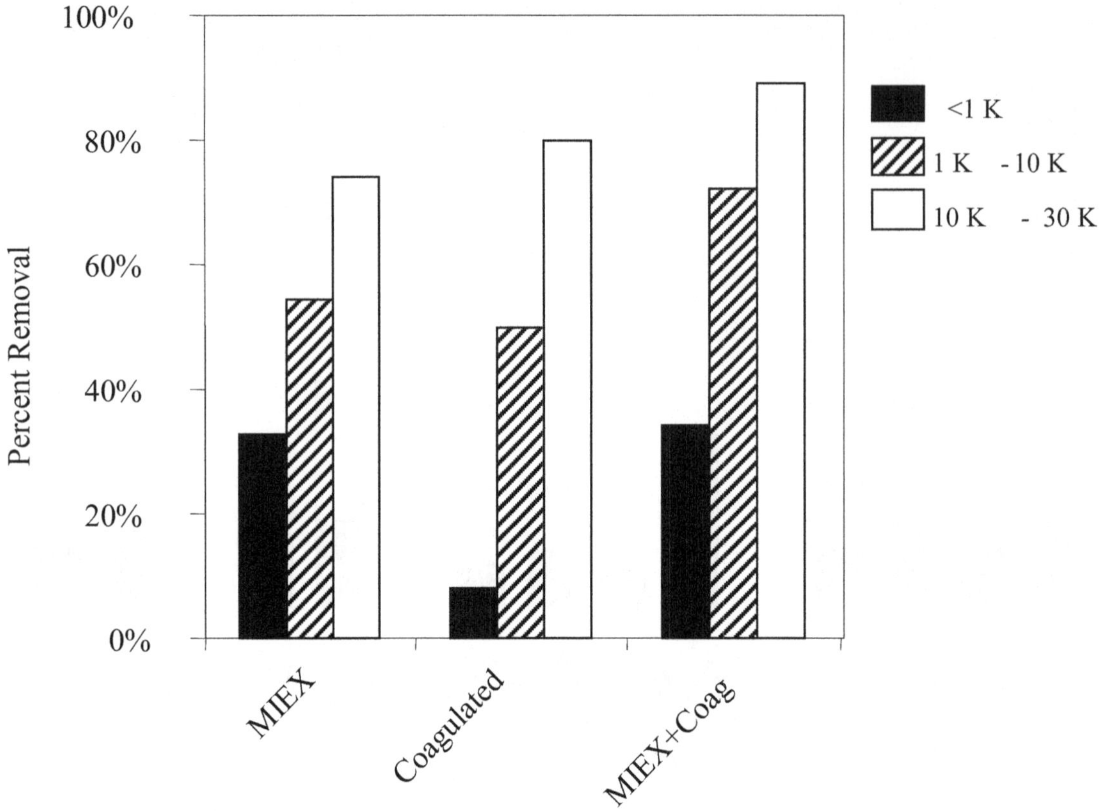

Figure 6.6 Effect of MIEX® and coagulation on the removal of different molecular weight fractions at HWTP

CHANGES IN DBP FORMATION POTENTIAL

NOM is the principal precursor of DBPs, and DBP formation is directly proportional to the concentration and characteristics of the NOM. Based on the decrease in DOC and UV-absorbing material observed by MIEX® treatment, a decrease in both THM and HAA formation potential was expected. Figures 6.7 and 6.8 reflect the THM and HAA formation potential after 3 days of contact with chlorine. The results shown are averages of Week 1 and Week 4 values and are shown in μmol/L units because of the differences in the molecular weight of the DBP species. The molar reduction of 3-day UFC THM formation potential was high at the HWTP (65 percent) and the FTWTP (63 percent), and lower at the MWTP (40 percent) and the AMSWTF (32 percent). The lowest removal at the AMSWTF is consistent with the low DOC and UV254 removals reported earlier. The removal of 3-day UFC HAA formation potential was also higher at the HWTP (73 percent), the FTWTP (68 percent), and the MWTP (64 percent), and lower at the AMSWTF (45 percent).

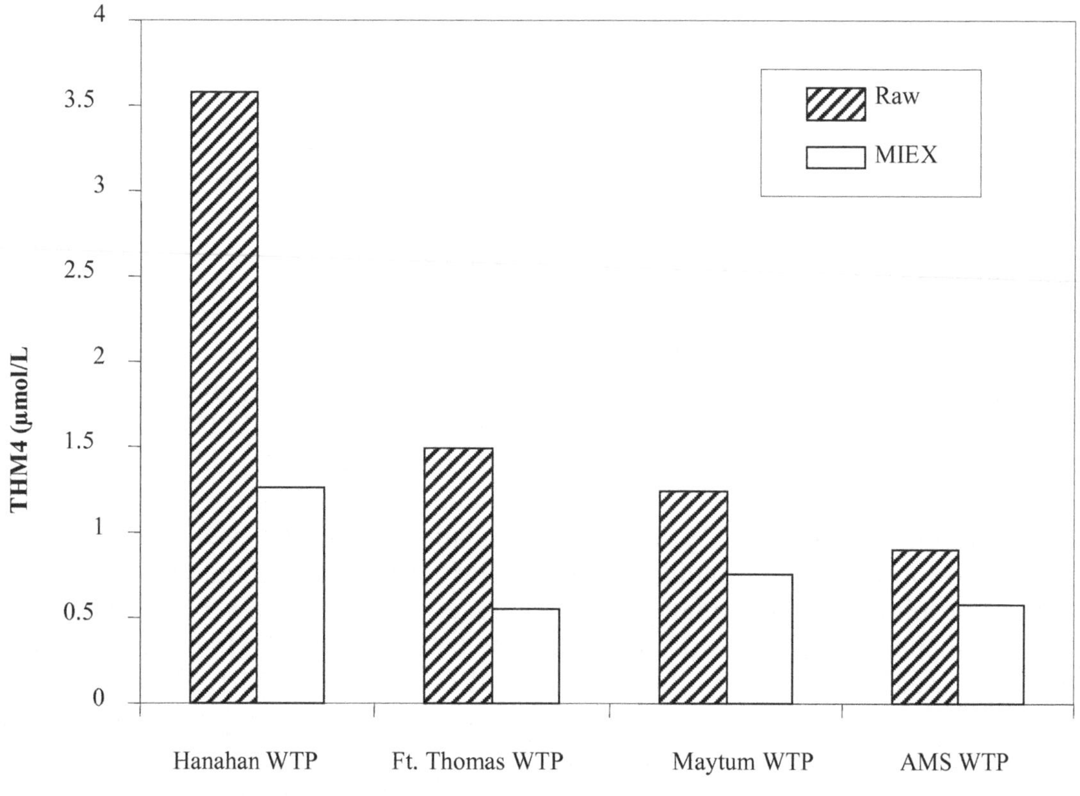

Figure 6.7 Average THM4 concentrations before and after MIEX® treatment after 3 days of contact with free chlorine

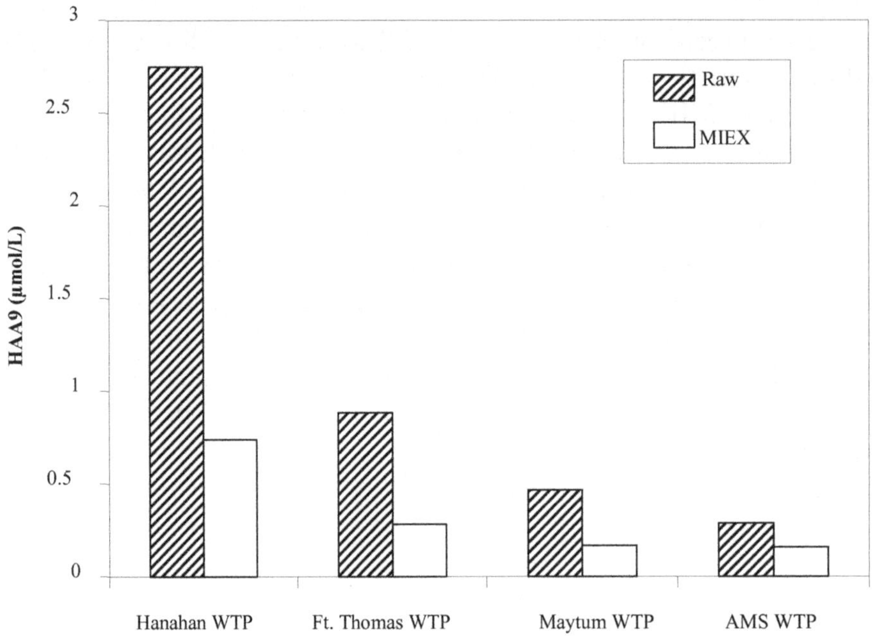

Figure 6.8 Average HAA9 concentrations before and after MIEX® treatment after 3 days of contact with free chlorine

The following data mirror the decreases in DOC and UV-absorbing material shown on Figures 6.1 and 6.2, where the removals of the two NOM surrogates were also extensive at the HWTP, the FTWTP, and the MWTP, and lowest at the AMSWTF. Figures 6.9 and 6.10 reflect the decrease in THM and HAA formation potential after MIEX® treatment, after coagulation, and after MIEX® treatment followed by coagulation at the HWTP and the FTWTP following 3 days of chlorine contact. Figure 6.9 shows that MIEX® treatment generally provided equivalent or slightly better removal of THM precursors than coagulation. One exception to this trend occurred during testing in Week 4 at the HWTP when coagulation alone achieved a slightly higher reduction in THM formation. Maximum removal of THM formation potential was achieved with MIEX® treatment followed by coagulation in all cases, even though a lower dose of coagulant was used. This supports the findings of Singer and Bilyk (2002) that THM and HAA formation potential is decreased by coagulation, but is decreased even more by MIEX® treatment. The removals of HAA precursors by MIEX® treatment, as compared to coagulation and to MIEX® treatment followed by coagulation show a similar pattern, as shown on Figure 6.10. These data mirror the removals of DOC and UV-absorbing materials shown in Figures 6.1 and 6.2.

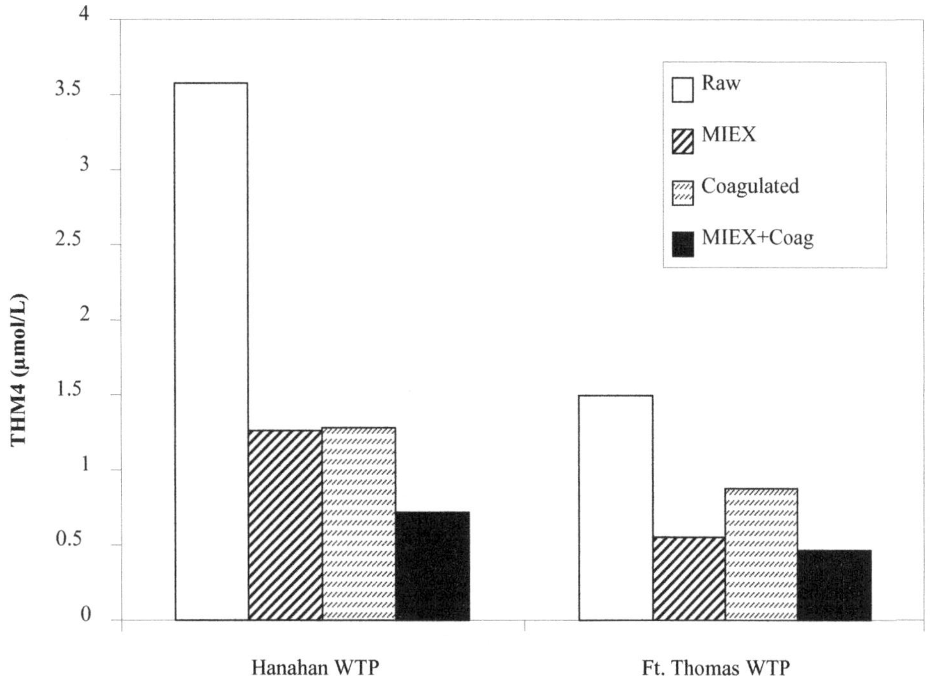

Figure 6.9 Average reduction in THM4 concentrations for various treatments after 3 days of contact with free chlorine

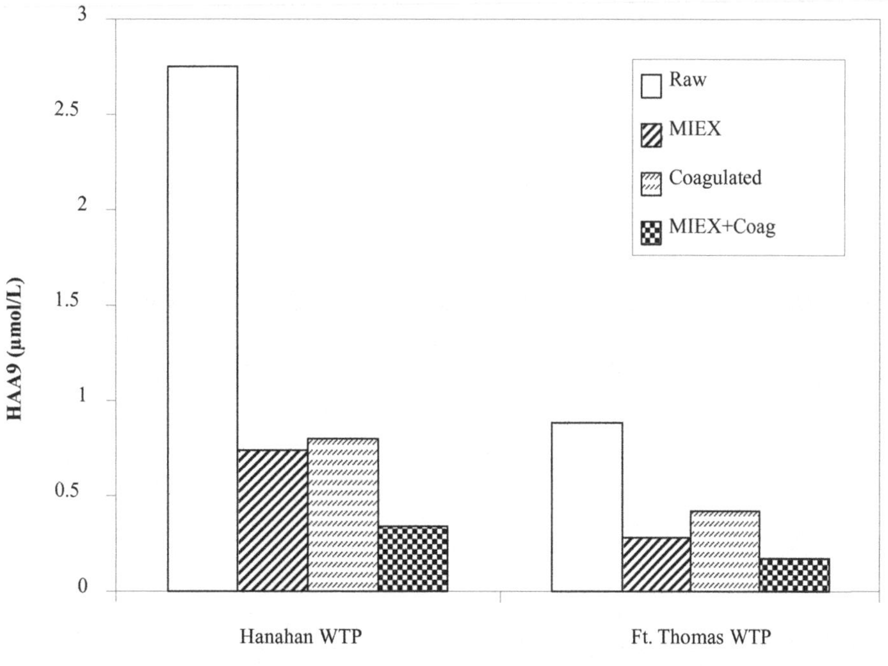

Figure 6.10 Average reduction in HAA9 concentrations for various treatments after 3 days of contact with free chlorine

Figure 6.11 illustrates how THM formation increased over time during testing conducted at the FTWTP. THM concentrations in each sample initially increased rapidly, and then proceeded more slowly. As expected, the highest THM formation potential occurred in raw water. By reducing the concentrations of DOC and UV-absorbing materials, MIEX® treatment and coagulation lowered the THM formation. The greatest reduction in THM formation occurred after treatment with MIEX® followed by coagulation, but the majority of the decrease was attributed to the MIEX® treatment.

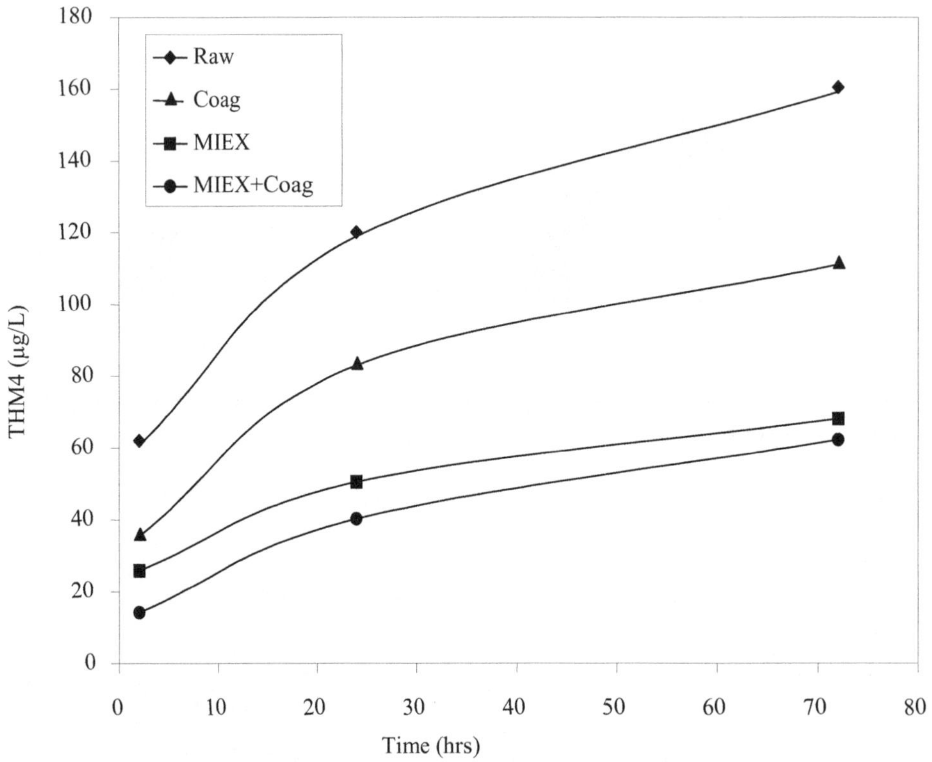

Figure 6.11 Fort Thomas WTP THM4 formation during Week 4 with various treatments

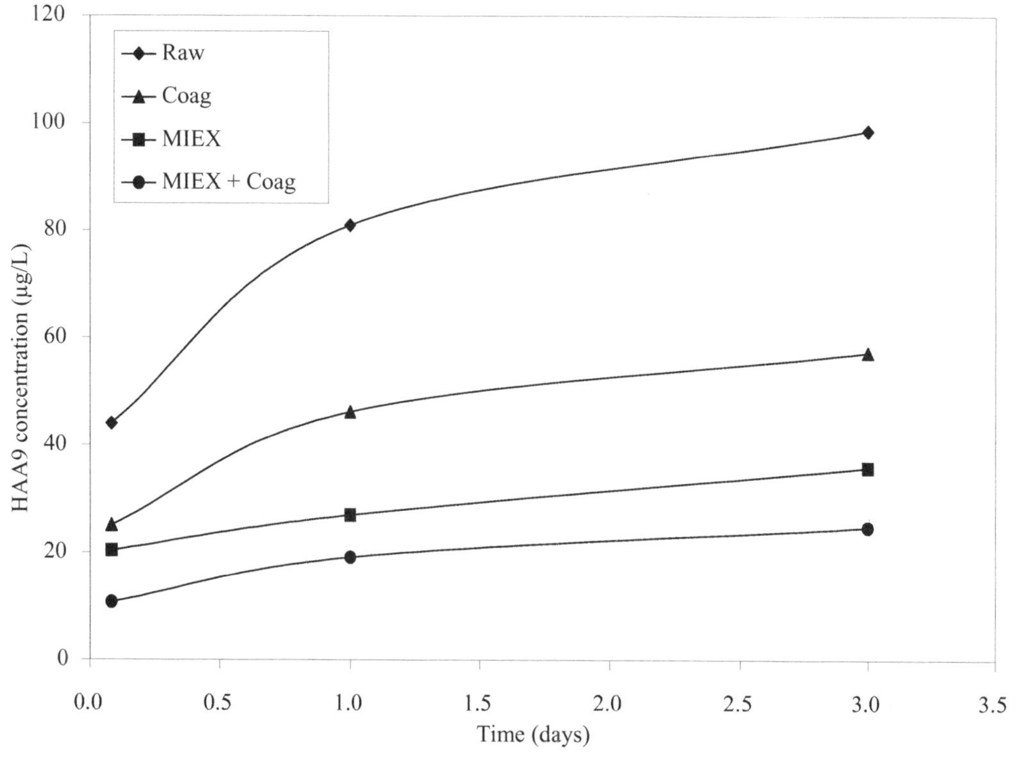

Figure 6.12 Fort Thomas HAA9 formation during Week 4 with various treatments

Figure 6.12 illustrates how HAA concentrations increased over time during testing conducted during week 4 at the FTWTP. HAA concentrations in each sample initially increased rapidly, and then proceeded more slowly. Similar to THM formation, the raw water sample exhibited the highest HAA formation potential. MIEX® treatment and coagulation lowered HAA formation, as expected from the reduction in DOC and UV-absorbing materials. The greatest decrease in HAA formation occurred with MIEX® treatment followed by coagulation, but the majority of the decrease was attributed to the MIEX® treatment.

Figure 6.13 summarizes the percent removal of THM formation potential for each utility based on the 3 day chlorination results. The results show that MIEX® treatment removed about 60 percent of the THM formation potential at HWTP, FTWTP, and MWTP. At the AMSWTF, removal of THM formation potential was only 30 percent, corresponding to the lower DOC removal.

Figure 6.14 indicates that the percent removal of HAA precursors was approximately 70 percent at the HWTP, FTWTP, and MWTP. This is approximately equivalent to the percent removal of UV-absorbing material at each location, as seen on Figure 6.2. Removal at AMSWTF was only 41 percent, consistent with the lower removal of UV-absorbing material.

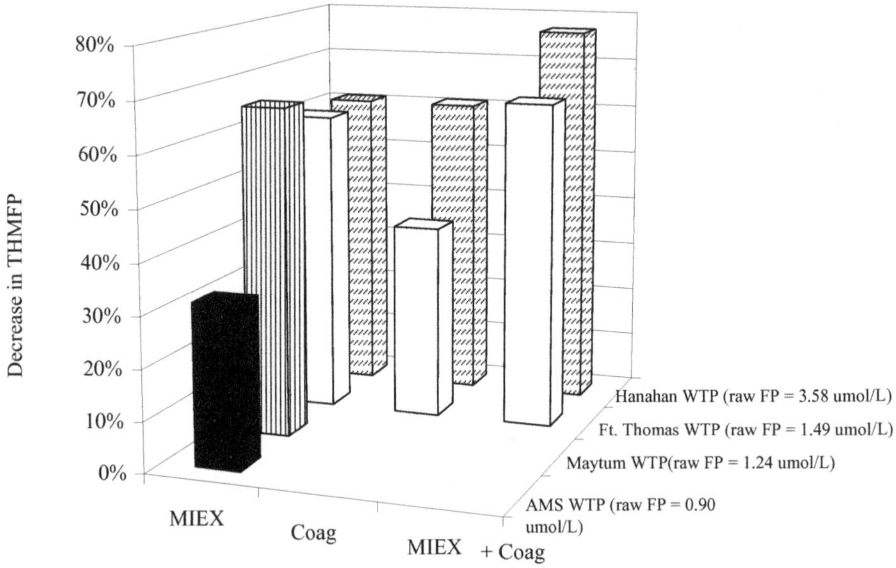

Figure 6.13 Average decrease in THM4 formation potential by MIEX®, by coagulation, and by MIEX® followed by coagulation

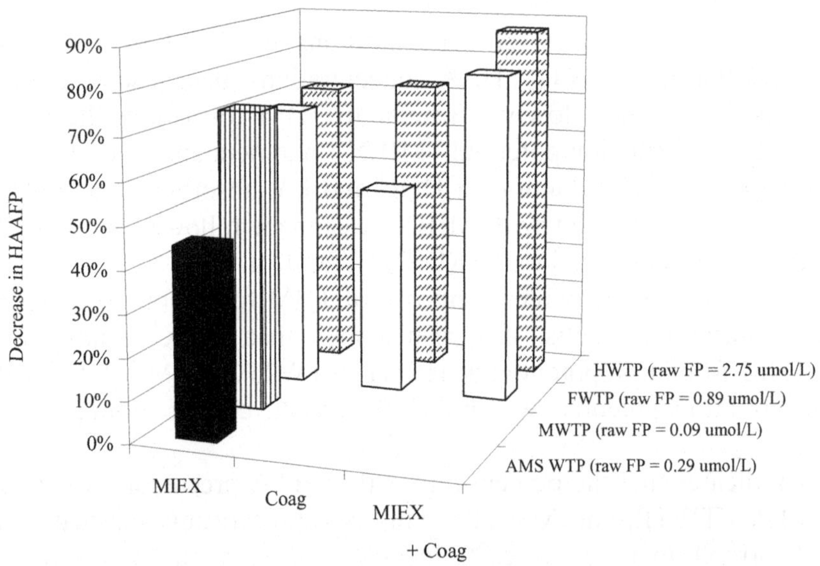

Figure 6.14 Average decrease in HAA9 formation potential by MIEX®, by coagulation, and by MIEX® followed by coagulation

EFFECTS OF MIEX® TREATMENT ON DBP SPECIATION

Testing results demonstrated that MIEX® treatment removes appreciable amounts of DOC and UV-absorbing material, and thus lowers the overall THM and HAA formation. Previous research has shown that reducing organic carbon concentrations prior to chlorination reduces total THM and HAA formation (Babcock and Singer 1979, Edzwald et al. 1985, Singer 1999), but that the concentration of some of the brominated THM and HAA species, which may be of greater health concern than the corresponding chlorinated species (Linder et al. 1997, Pegram et al. 1997, Kroll et al. 1994), can increase in waters containing bromide. Concentrations of brominated THMs and HAAs formed during chlorination can increase as the organic carbon concentration of the water decreases, because lower organic carbon concentrations require a lower chlorine dosage to achieve the desired chlorine residuals. Therefore, the bromide-to-chlorine ratio increases with decreasing organic carbon concentration, thus shifting the distribution of THMs and HAAs to the brominated forms (Symons et al. 1993, Cowman and Singer 1996, Black et al. 1996).

Figure 6.15 shows the THM speciation of chlorinated raw water from the HWTP for Week 1. Chloroform was the dominant species and the brominated species comprised less than 15 percent of the total THMs. Figure 6.16 shows the corresponding speciation of THMs after MIEX® treatment. Total THM formation decreased from 500 μg/L to 168 μg/L (66 percent reduction). Chloroform concentrations dropped from 415 μg/L to 86 μg/L (79 percent reduction), and $CHBrCl_2$ levels decreased from 78 μg/L to 48 μg/L (38 percent reduction), whereas $CHBr_2Cl$ increased from 6.5 μg/L to 25 μg/L. Similar findings were observed at HWTP during Week 4, and at the other locations as well.

The effects of MIEX® treatment, coagulation, and MIEX® treatment followed by coagulation on the concentrations of tri-halogenated acetic acid (X_3AA) in the HWTP water during Week 1 are shown on Figure 6.17. The X_3AA concentration was highest in the raw water. MIEX® treatment and coagulation reduced the concentration of X_3AA to approximately equal degrees, and the largest decrease in X_3AA was observed when MIEX® treatment and coagulation were used together, although the increased removal is attributable primarily to MIEX® treatment. Similar results were achieved for di-halogenated acetic acids (X_2AA) at all locations. Total X_2AA concentrations at HWTP during Week 1 are shown on Figure 6.18.

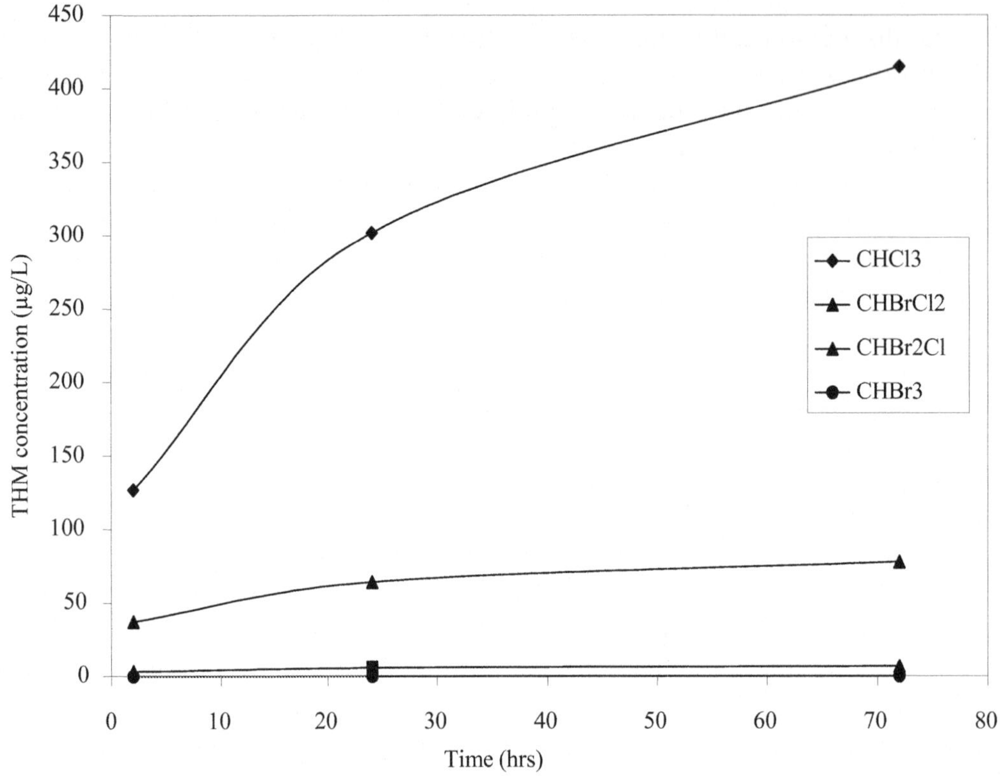

Figure 6.15 THM speciation for HWTP raw water during Week 1

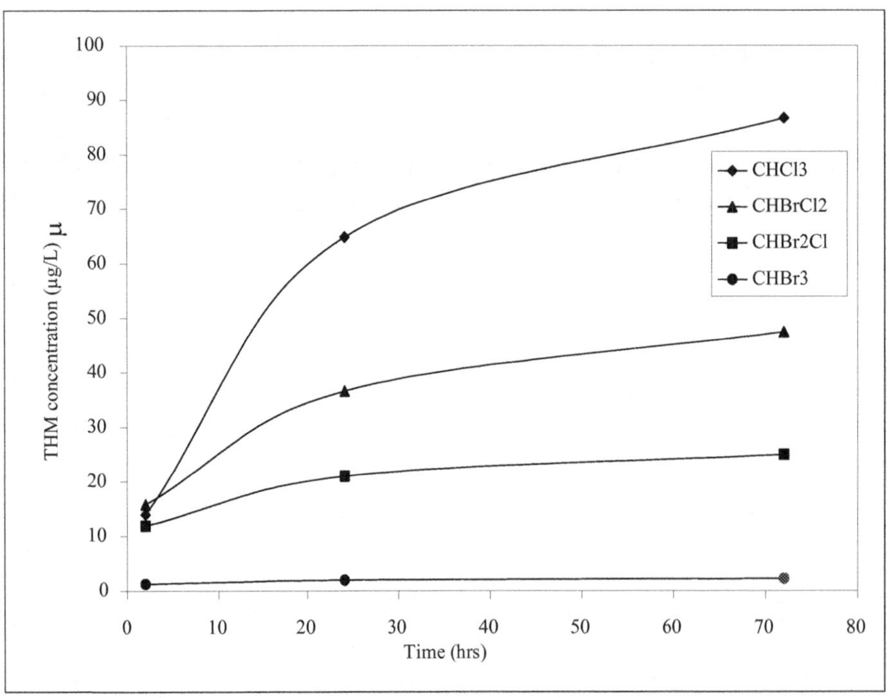

Figure 6.16 THM speciation for MIEX®-treated HWTP water during Week 1

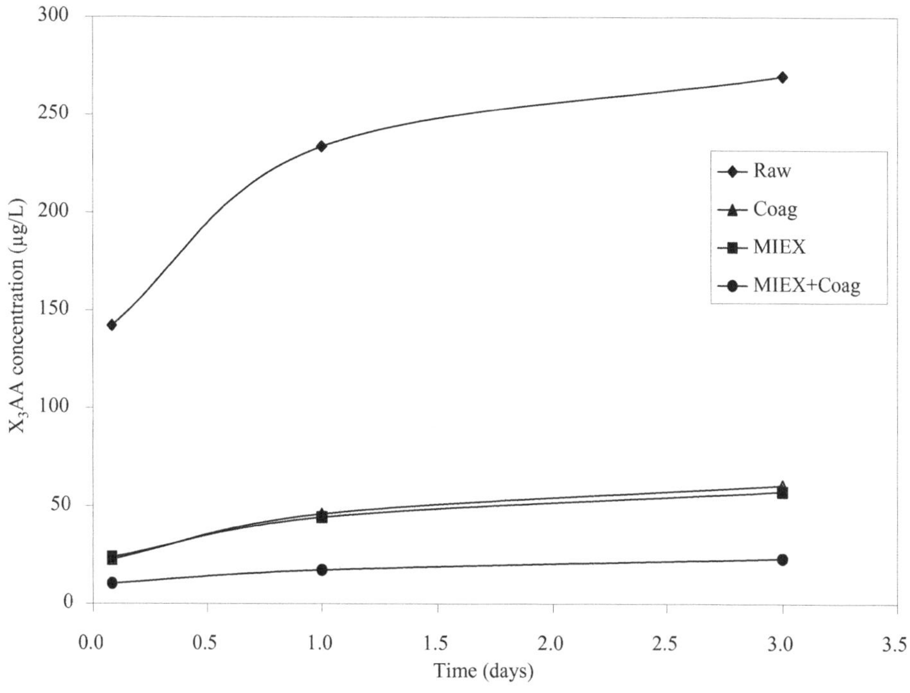

Figure 6.17 Total trihaloacetic concentrations for HWTP during Week 1

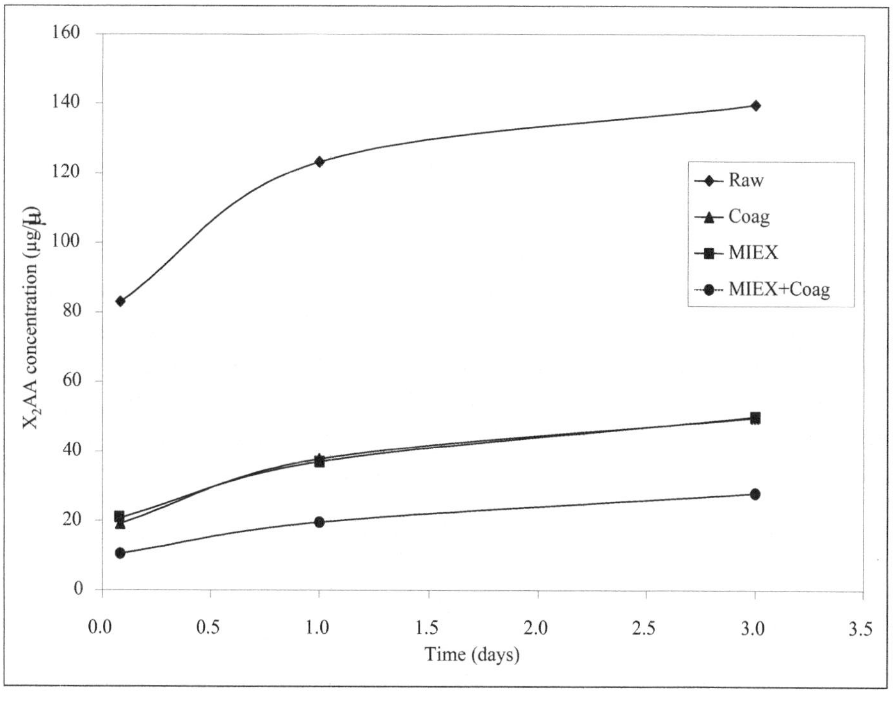

Figure 6.18 Total dihaloacetic concentrations for HWTP during Week 1

The bromine incorporation factor (BIF) shows the change in bromine incorporation into DBPs before and after treatment by the various processes. It represents the average number of moles of bromine in the THMs (or HAAs) per mole of THM4 (or HAA9). It is calculated by dividing the sum of the molar concentration of bromine in each individual species by the total molar concentration of all species, and is used to quantitatively describe the degree of bromine substitution in the THM distribution (Ko et al. 1996, Chang et al. 2001, Miles 2002).

Table 6.1 shows the changes in BIF for the THM4 species for each treatment process vs. time. The BIF values indicate that bromine is quickly incorporated into the THM species and as time progresses, the extent of bromine incorporation decreases. This suggests that the bromine-containing THM species are formed more rapidly than chloroform, which is consistent with previous research (Ko et al. 1996, Chang et al. 2001, Miles 2002). Additionally, the BIF increases as a result of treatment, consistent with the results shown on Figures 6.15 and 6.16, and the observation that less DOC was available in proportion to the amount of bromide. Similar results were obtained at the HWTP during Week 4, and at the other utilities.

Table 6.1
Changes in THM concentration and bromine incoporation factor for HWTP - Week 1

	Time	THM4 (μg/L)	BIF
Raw	2 hours	168	0.20
	1 day	372	0.15
	3 days	500	0.14
MIEX	2 hours	41.7	0.82
	1 day	124	0.51
	3 days	161	0.49
Coagulation	2 hours	56.5	0.67
	1 day	141	0.49
	3 days	192	0.40
MIEX + coagulation	2 hours	29.8	0.95
	1 day	83.6	0.80
	3 days	119	0.65

Changes in BIF for the HAA9 species for each treatment process vs. time are shown in Table 6.2. The increase in BIF from the raw water to the MIEX®-treated water and the coagulated water indicates a shift to the more highly brominated species as a result of treatment, which is consistent with the results presented on Figures 6.17 and 6.18 and with the finding that less DOC was available in proportion to bromide. As in the case of THMs, the rate of incorporation of bromine into the HAAs was initially very rapid, and then decreased with time.

Table 6.2
Changes in HAA9 concentration and bromine incoporation factor for HWTP - Week 1

	Time	HAA9 (μg/L)	BIF
Raw	2 hours	231	0.15
	1 day	368	0.05
	3 days	417	0.03
MIEX	2 hours	49.4	0.97
	1 day	88.4	0.58
	3 days	112	0.42
Coagulation	2 hours	49.0	0.73
	1 day	91.0	0.50
	3 days	112	0.28
MIEX + Coagulation	2 hours	31.2	0.60
	1 day	45.1	0.48
	3 days	45.3	0.38

The relative rate of incorporation of bromine into THMs and HAAs for the same water is also of interest. Obolensky and Singer (2003) have found a strong correlation between the percent of bromine in THMs and the percent of bromine in the di- and tri-HAAs. Figure 6.19 shows a comparison of the BIF for both THM4 and X_2AA in all four waters for all of the treatment processes evaluated. The incorporation of bromine into THMs is strongly correlated with the incorporation of bromine into dihaloacetic acids. The slope of the regression line is not 1:1, as found by Obolensky and Singer (2003), but there is a strong correlation nonetheless and the difference in the relationship may be attributable to differences in pH and temperature. Liang and Singer (2003) have shown that at pH 8.0, the mole fraction of brominated species favors the formation of THMs over HAAs, which is consistent with the results shown on Figure 6.19.

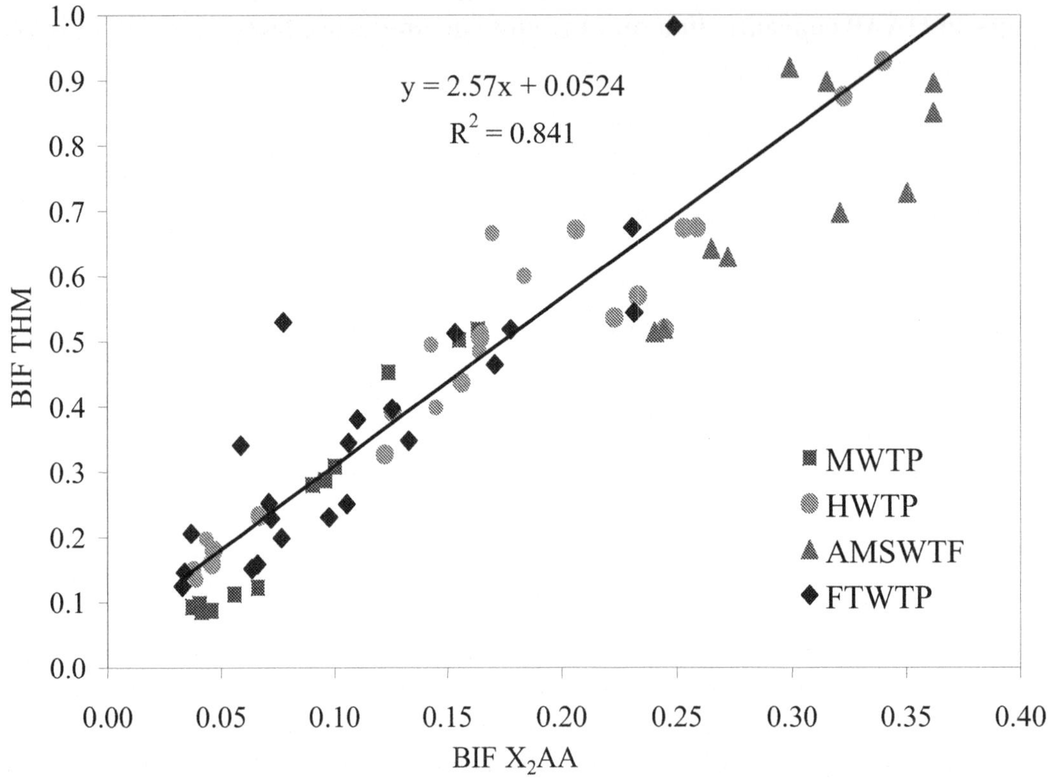

Figure 6.19 Comparison of bromine incorporation factor in THMs and dihaloacetic acids at all testing locations

The relationship between the percentage of bromine in THMs and X_3AA is shown on Figure 6.20. The relationship showed more scatter. There was similarity in the results from the MWTP and the FTWTP, which differed from those shown for the AMSWTF. The values from the HWTP did correlate well and highly favored incorporation into the THMs. The uptake of bromide into THMs and HAAs appears to exceed the raw water concentrations and the selectivity of the MIEX® resin for bromide needs to be further examined. The R^2 value is based on data from MWTP and FTWTP.

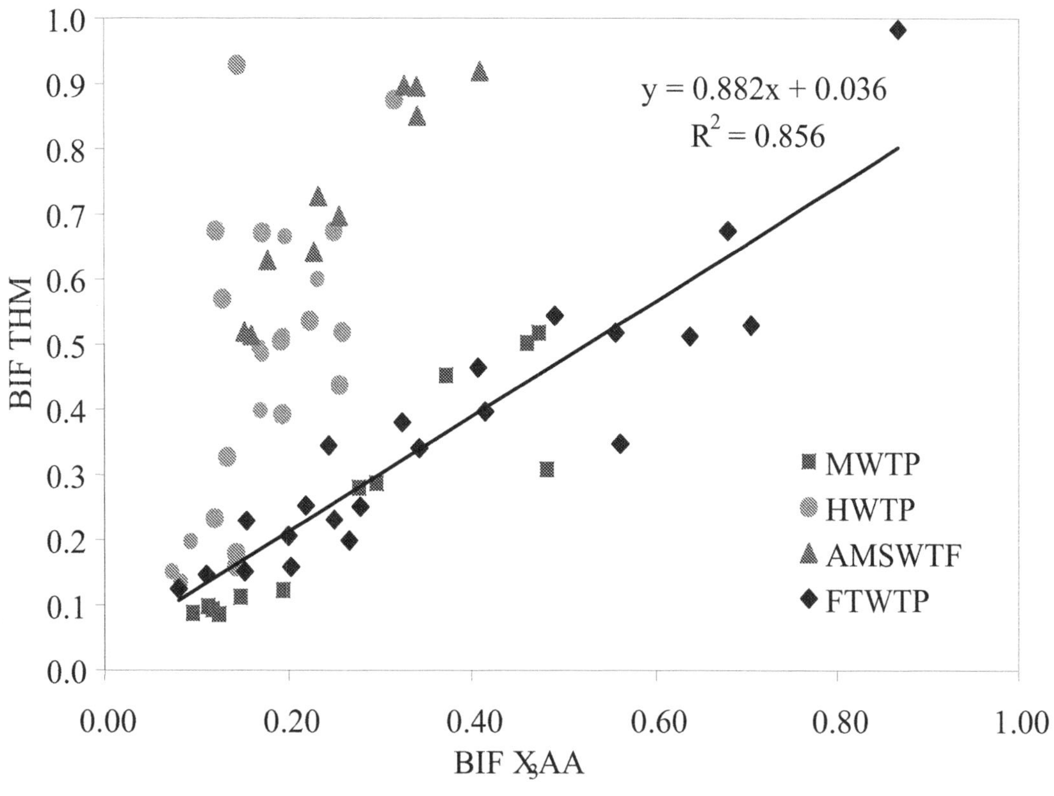

Figure 6.20 Comparison of bromine incorporation factor in THMs and trihaloacetic acids at all testing locations

CHAPTER 7
DISCUSSION AND COST EVALUATION

NATURE OF THE MIEX® PROCESS

The MIEX® process is an ion exchange process that employs a strong base anion exchange resin to remove NOM from water. NOM removal occurs as a consequence of the negative charge characteristics of many NOM components typically found in water. A resulting attraction for anion exchange sites occurs where these components are sorbed in exchange for chloride, which is released. The strong base properties of the anion exchange resin allow for effective performance over a wide range of pH. There is a spectrum of properties among the various NOM fractions, and there is a variation in the response of different fractions to removal.

Magnetized iron is incorporated into the MIEX® resin, thereby allowing the resin particles to be attracted to one another without adding coagulants. This attraction is overcome by mixing that keeps the resin in suspension during the contacting phase, while manifesting an attractive force between the resin particles in the more quiescent settling phase to result in effective capture of the resin for recycle. As a consequence, the MIEX® process can be operated in mixed basins followed by recovery of the resin in specially designed sedimentation basins.

Testing at the four sites evaluated during this study indicated a variation in the susceptibility of NOM to removal by the MIEX® process. Higher levels of DOC removal were observed at the Maytum WTP and the Hanahan WTP than at the other two locations. Removal of DOC at these locations approached 60 percent or more. In contrast, DOC removals at the Fort Thomas WTP were slightly lower (50 to 60 percent), and removals of less than 35 percent were observed at the Alfred Merritt Smith WTP. Therefore, differences in performance can be expected from one location to the next, emphasizing the need for site specific evaluation.

NOM fractionation evaluations at all four sites indicated a preference of the MIEX® resin for the highest molecular weight and most hydrophobic fractions, followed in order of preference by the intermediate molecular weight and transphilic fractions. The lowest affinity was for the lowest molecular weight and hydrophilic fractions. This observation may be unique to these study sites as others (Lee et al. 2003) have reported a high degree of affinity for low molecular weight fractions.

Evaluations of the effects of MIEX® and coagulation processes in combination were performed at the HWTP and the FTWTP. At HWTP, a significant increase in the removal of NOM was observed when the two processes were operated in combination, thereby confirming the significance of complementary removal at some locations. This effect was less pronounced at the FTWTP. Therefore, the extent of complementary action may be site specific

Removal of Inorganic Anions

The anion exchange properties of the MIEX® resin suggest that removal of several inorganic anions of interest may be possible. Observations in this study suggest that the effectiveness may vary depending on several conditions. At MWTP, sulfide was very effectively removed throughout the entire study period. This removal may have been augmented by biological activity, as significant growth of filamentous bacteria was observed, and progressive improvements in sulfide removal occurred as these bacteria became established.

Removal of bromide was highly variable. Effective removal was established at the HWTP when rock salt rather than sea salt was used for preparing the regenerant brine. This change was made due to the presence of significant levels of bromide in sea salt and the potential that this creates for interference with regeneration of the resin with respect to bromide uptake. While changing the salt source was very effective in improving bromide removal at the HWTP, it was not effective at the AMSWTF where TDS in the source water may have been competing with bromide for ion exchange sites. In particular, the high concentrations of sulfate could have been a factor at this location, as sulfate is a divalent anion that has a much higher affinity for anion exchange sites than a monovalent anion such as bromide.

The removal of arsenate and perchlorate was also evaluated at the AMSWTF. Bench-scale tests with spiked samples resulted in arsenic removals up to 40 percent and perchlorate removals up to 70 percent under conditions that simulated a MIEX® concentration up to 20 mL/L, a detention time up to 20 minutes, and a 5 percent regeneration rate. Higher removals might be possible at higher resin concentrations or at higher regeneration percentages. High levels of competing anions in the raw water at AMSWTF also are likely to have reduced performance. Raw water concentrations of arsenic and perchlorate were lower in the pilot plant evaluations and the percent removals observed for this phase of evaluation were correspondingly lower. Arsenic levels in the raw water were at levels that were too low to assess, while a small degree of perchlorate removal was observed at the low levels that were present (4 to 14 μg/L).

While the removal of arsenic and perchlorate is beneficial from the standpoint of contaminant removal, high concentrations of these anions in the brine could determine whether the brine may need to disposed under a hazardous categorization. This will be somewhat site-specific, depending on the concentration of the contaminants and on local regulations.

Effect of the MIEX® Process on General Water Quality Characteristics

One outcome of the anion exchange process is the increase in chloride concentrations that occurs as a consequence of exchange between the chloride at the resin surface and the other anions that are being sorbed. Results of this study indicate that chloride concentrations typically increase by 20 to 40 mg/L. Adding a final rinse step to remove excess salt from the resin during regeneration reduced the degree of increase in chloride concentrations in this case. Higher chloride levels could occur if the rinse step is eliminated. Therefore, some variations in finished water chloride concentrations would be expected for full-scale installations, depending on whether a resin rinse step is incorporated.

Factors that Affect Performance

The MIEX® process does not involve direct application of additives such as oxidants, coagulant chemicals, or powdered activated carbon. Rather it is a recycling-based process which is initially charged with a specific concentration of resin, and only a portion of the resin is regenerated in each operating cycle, and thus most of the resin is used several times within the process before it is regenerated. As a result, the resin loses some of its original capacity, but maintains a lower steady-state capacity through several cycles. The level of the steady-state capacity depends largely on the regeneration percentage. This distinction is important with respect to the selection of the system's operating conditions. In this study, the resin condition was simulated in bench-scale evaluations by using the pilot plant as a source for both freshly regenerated resin and un-regenerated recycled resin that were blended in proportions to simulate the desired operating conditions. This approach yielded bench-scale results that provided a reasonable prediction of the types of results that were subsequently achieved in pilot testing.

Resin concentration and contact time also have an effect on process performance. Evaluations at the full-scale Wanneroo Groundwater Treatment Plant (WGTP) indicated a tradeoff between resin concentration and contact time. This tradeoff makes it possible to reduce the investment in contactors operating at a higher resin concentration. This may increase the volume of some regeneration components to accommodate the higher resin concentration, but will not necessarily increase the resin use, as this is related to the rate of loss as opposed to the resin concentration that is maintained in the system. Recent observations at the WGTP suggest that resin loss may actually be reduced as the concentration is increased, thereby reducing the operation costs as compared with operation at lower concentrations. Resin concentrations less than 10 mL/L were therefore not considered appropriate for pilot testing in this study. This effect is a direct illustration of the distinction between processes involving the application of additives whose concentration will be dictated by the quantity added to the system, and the application of the MIEX® resin in a recycle system.

Interaction with Other Processes

NOM affects other plant processes, and its removal by the MIEX® process at the head of a plant can yield significant overall improvements in finished water quality. Evaluations at the HWTP demonstrated that the MIEX® process had a significant effect on coagulation by making it possible to achieve effective coagulation over a broad range of pH and alum doses, allowing the alum dose to be reduced from 55 mg/L without MIEX® treatment to as low as 10 mg/L following MIEX® pretreatment. The increased range of effective coagulation conditions when using MIEX® treatment also would reduce the need for precise process control at this location, a consideration that would improve the overall stability of operation. This reduction would have the added benefit of reducing the quantity of solids produced by the coagulation process and the associated disposal costs. This level of reduction in coagulant dose could be a factor that would allow a broader range of clarification alternatives and/or lower oxidant/disinfectant demands. Both considerations are factors that could affect facility designs of other components of the treatment plant in ways that could reduce the capital cost impact of adding a MIEX® system is some cases.

Evaluations at the FTWTP were conducted using a mixed coagulant sequence which included addition of 5 mg/L of ferric sulfate along with addition of a commercial acidified alum-polymer blend at dosages that were varied in response to changes in raw water quality conditions. Adding MIEX® treatment in this sequence resulted in a shift in optimum coagulation conditions, but did not make it possible to reduce the dose of the acidified alum blend. Subsequent evaluations indicated that MIEX® treatment made it possible to use low doses of ferric sulfate without adding any acidified alum. Without MIEX® treatment, the coagulant demand for ferric sulfate was much higher. Therefore, the use of MIEX® treatment would simplify the coagulation process at this location as well as reducing the coagulant dose.

While the MIEX® process may significantly benefit systems where a clarification step is employed prior to granular media filtration, the potential benefits for direct filtration systems are less definitive, as the MIEX® process causes an increase in particulate levels across the process. In systems that include a clarification step, the reduction in NOM offsets this adverse effect, and removal of particulates during clarification becomes more effective. However, direct filtration is not preceded by an intermediate particle removal step. Evaluation of a MIEX®-direct filtration sequence at AMSWTF indicated that the removal of NOM did not offset the effect of the increase in the particulate load to the filters, and the rate of head loss buildup was higher when the MIEX® process was applied as a pretreatment process.

The effect of NOM removal by the MIEX® process on downstream oxidation processes at all test locations was a substantial reduction in chlorine demand. Evaluations at AMSWTF indicated that there was an initial increase in the ozone demand following MIEX® treatment; however, the rate of ozone decay was much slower, and it is possible that the increase in demand was associated with carryover of resin from the MIEX® process. Therefore, the presence of a clarification step between the MIEX® process and ozonation may have a different effect on initial demand. It is also possible that the MIEX®–ozonation sequence would be of greater benefit where DOC concentrations in the raw water are higher and are associated with higher removal percentages than at the AMSWTF.

Full Scale Experience with the MIEX® Process

The Wanneroo Groundwater Treatment Plant (WGTP) in Western Australia has used the MIEX® process at a design capacity of 112.5 ML/day (30 MGD) for more than two years. The groundwater supply contains DOC at concentrations above 10 mg/L as well as high levels of iron. Oxidation of iron upstream from the MIEX® process has potential to cause fouling of the resin as a result of iron precipitation. However, this has not prevented successful operation of the MIEX® process at this location.

This facility has generally met treatment goals, which confirms the capability for operating the process at full scale. Key design and operational aspects for this plant are:

- Contact time.
- Resin concentration.
- Regeneration rate.
- Sedimentation overflow rate.

The MIEX® process at WGTP is operated at a contact time of 10 minutes, rather than the 30 minutes used in the original concept design, a change that was facilitated by a correspondingly higher resin concentration. This modification proved effective in pilot testing, and has worked well in the full scale operation of the facility. Regeneration rate has proved to be an important factor in meeting the goals for NOM removal, which may be attributable to the high concentrations of DOC in the source water as compared to the quantity of resin in the system. This level of sensitivity was not observed at any of the four test sites in this study, where the DOC concentrations in the source water were lower.

While the MIEX® process at WGTP was designed to operate at a resin concentration of 20 mL/L, the concentration was recently increased to 35 mL/L to reduce resin loss by improving the opportunity for contact between resin particles in the settler inlet wells at the higher concentration.

An important consideration in the design and operation of the MIEX® system is the need to minimize exposure of the resin to excess shear forces that can lead to physical damage. There is also a need to handle the resin as a slurry. The following design features were incorporated into the system to deal with these issues:

- Low shear hydrofoil mixers operated at low speed were installed in the contact basins, and low shear pumps operated at a relatively slow speed are used to minimize physical damage to the resin.

- The sides of the hoppers in the settling tanks were constructed at a 70 degree angle to convey the resin to a central collection point without the use of mechanical scrapers or collection devices that would cause physical abrasion or plugging by the accumulated slurry.

- Adequate velocity is maintained in resin transport piping to keep the resin in suspension and to avoid plugging. Provisions are made for flushing piping segments before they are taken out of service or when a pump is taken off-line. Long radius bends are used in the piping to minimize resin attrition due to abrasion and also to minimize the tendency for plugging.

Disposal of Brine

Disposal of waste brine is a key issue that will affect the overall viability of the MIEX® process at a given location. WGTP developed a disposal option utilizing an ocean outfall that was acceptable to the regulatory community in Western Australia. Dilution factors of up to 1000 were projected for this system. Brine is transported by truck from the inland location of the WGTP to the disposal location.

The quantity of waste brine generated in the MIEX® process will vary depending on the regeneration percentage and the resin concentration. The use of a rinse step during regeneration will add to the volume of wastewater to be disposed, and becomes a decision that will have a significant effect on the volume of waste brine. In general, this volume could be relatively small (2,500 to 6,000 gallons per each 10 million gallons of water treated). While ocean disposal has proved to be feasible at the WGTP, there are other options available, such as discharge to wastewater treatment facilities via the municipal collection system, mechanical drying to further reduce the volume to be disposed, solar drying in lagoons where climate is appropriate, and treatment for removal of DOC from the waste brine to allow reuse.

Each location will have a site specific set of issues relative to waste disposal and requirements for compliance with states and local regulations. Disposal in some areas may be affected by the presence of anionic contaminants that may be removed by the MIEX® process and concentrated in the regenerant brine.

COST EVALUATION

Experience at the WGTP was applied to develop opinions of probable costs for US construction of a MIEX® system designed for a peak treatment capacity of 37.5 ML/day (10 mgd). An operational resin dose of 20 mL/L and a contact time of 15 minutes were selected as the basis for the cost opinion. The settling tanks were sized for a loading rate of 6 gpm/sq ft. It is possible that costs could be reduced in the future if refinements are made over time.

An itemized capital cost opinion for construction of the facilities required to provide a functional MIEX® process are as follows:

General Conditions/Site Preparation Work/Site Services/Buried Piping	$160,000
Contactors	$720,000
Settling Tanks	$330,000
Resin/Resin Handling/Chemical Storage and Handling	$280,000
Process Pumps and Above Ground Piping	$560,000
Electrical, Instrumentation and Control Instrumentation	$580,000
Engineering, Start-up, Commissioning, Project Management	$460,000
Construction Overhead and Profit	$410,000
Licensing Fee	$40,000
Subtotal	$3,540,000
Cost of Initial Quantity of MIEX® Resin	$330,000
Total Probable Capital Cost	$3,870,000

Operation and maintenance costs were developed based on a resin cost of $10 per liter of settled resin. Projected costs were developed for two levels of resin loss. One level of loss is based on full-scale operating experience at the WGTP. This loss rate is estimated to be 3.0 liters of resin per million liters of water treated. Evaluations at the WGTP and by others have been directed towards further reduction in this rate of loss and new concepts for capture of the resin by settling and other means are being examined. A second opinion of probable operation and maintenance costs based on a loss rate of 2.0 liters of resin per million liters was developed to provide insight into the economic benefit associated with reducing resin losses.

An average annual flow of 25.1 ML/day (6.7 mgd) was used in estimating annual O&M costs. Capital costs are annualized through amortization at an interest rate of 6 percent over a period of 20 years. These costs are compiled below for the two conditions of resin loss.

Case 1 – Loss of 3.0 liters of resin per million liters of water treated

Operation and Maintenance	
Annual cost of resin	\$270,000
Other annual treatment costs (salt, power, personnel etc.)	\$ 70,000
Total Annual Operation and Maintenance Cost	\$340,000
Operating and Maintenance Cost per 1000 Gallons Treated	\$0.14
Amortized Capital Cost	\$340,000
Total Annual Cost	\$680,000
Total Cost per 1000 Gallons Treated	\$0.28

Case 2 - Loss of 2.0 liters of resin per million liters of water treated

Operation and Maintenance	
Annual cost of resin	\$180,000
Other annual treatment costs (Salt, power, personnel, etc.)	\$ 70,000
Total Annual Operation and Maintenance Cost	\$250,000
Operating and Maintenance Cost per 1000 Gallons Treated	\$0.10
Amortized Capital Cost	\$340,000
Total Annual Cost	\$590,000
Total Cost per 1000 Gallons Treated	\$0.24

These costs do not include an allowance for disposal of waste brine. This cost can vary from a relatively modest amount if direct discharge to a sanitary sewer is allowed, to a substantial amount depending on the method of transport and disposal required. In addition to these considerations, the volume to be disposed will be affected by a decision of whether to provide a water rinse step for the resin at the end of regeneration as this will add to the volume to be disposed. Capital costs will be affected to an extent by this decision as adjustments for regeneration time and configuration will need to be considered. The decision regarding rinsing is affected by the impact that remnant brine from regeneration will have on sodium and chloride levels, and in terms of possible recycle of a limited quantity previously removed contaminants. Such decisions are site specific.

Cost estimates were based on minimal loss of resin capacity over time. There was little indication of progressive loss of resin capacity over the life of the MIEX® process at the WGTP in spite of the fact that the process is challenged by potential fouling resulting from upstream oxidation of significant levels of iron. Evaluations of resin activity over the 5 week periods of pilot plant operation at each of the sites in this study also did not exhibit differences over time.

While these observations were based on relatively short periods of resin life, they did tend to support the observed stability of capacity that has been observed at the WGTP. These observations are not sufficient for drawing definitive conclusions; more information will become available as full scale applications expand and it is still recognized that capability to maintain capacity may prove to be site specific. However, the resin shows promise for maintaining capacity based on information that is available to date.

Potential Applicability of the MIEX® Process

The potential applicability of the MIEX® process at each of the test locations is discussed below:

- Maytum WTP – The DOC concentrations in the raw water at the Maytum WTP averaged 1.93 mg/L at the time of testing. Although this level of DOC is relatively low, formation of THMs and HAA5 could exceed the limits of 80 μg/L and 60 μg/L, respectively, after contact with free chlorine for several days in the distribution system if the present treatment system is not augmented with a process for removing DOC. Use of the MIEX® process resulted in significant reductions in the DOC, with a corresponding reduction in THM and HAA5 formation that would more readily meet the respective regulatory limits. MIEX® treatment also reduced the concentration of sulfides in the raw water from an average of 1.7 mg/L to less than 0.02 mg/L. While removal of sulfide was high throughout the test period, biological activity may have supplemented the removal by the resin in the latter portion of the test period. Both of these treatment outcomes would provide a significant benefit at this location. Implementation of the MIEX process at this location would require the addition of a filtration step for removing resin fines that escape the process and contribute to turbidity. Such filtration could consist of conventional granular media filters or other filtering approaches that could include microstrainers (Rowland 2004).

- Hanahan WTP – The DOC concentrations in the raw water at the Hanahan WTP averaged 5 mg/L at the time of testing. DOC and UV254 reductions achieved by the MIEX® process were relatively high, particularly when it was followed by alum coagulation. THM and HAA5 formation following coagulation alone remained relatively high despite the relatively high alum dosage (55 mg/L) and the relatively low coagulation pH (5.8 to 6.0). The combination of MIEX® treatment followed by alum coagulation yielded significant reductions in the formation of both THMs and HAA5; however, THM formation still exceeded the limit of 80 μg/L under the test conditions. Process modifications that could further reduce THM formation include adjustments in coagulation, such as the use of alternative coagulants; higher regeneration rates; and higher resin doses.

  A significant benefit to coagulation was observed at this location that would allow effective reductions in alum dose from 55 mg/L down to as low as 10 mg/L, which would represent considerable saving in the cost for coagulant as well as reducing the quantity of residuals produced and the associated disposal cost. It would probably also eliminate the need for the coagulant aid polymer that is presently used in the absence of MIEX® treatment. Estimates of the savings in coagulant and for cost of disposal of solids at a landfill are approximately \$0.03/1000 gallons treated. This savings is compared to estimated MIEX® annual operation and maintenance costs of \$0.10 and \$0.14 that correspond to resin loss rates of 2.0 and 3.0 liters per million liters, respectively. This would reduce the net annual operation and maintenance costs to \$0.07 and \$0.11 per 1000 gallons for the respective levels of resin usage.

Evaluations indicated that significant reductions in bromide are possible at HWTP if the regenerant brine is prepared using a low bromide salt. Since high concentrations of bromide can periodically cause problems with disinfection by-product formation at this location, this capability would be particularly beneficial. Bromide removal at this location was probably also be facilitated by the presence of low levels of competing anions as removal of bromide proved to be more difficult when using a low bromide salt for brine makeup in testing at the AMSWTF where competing anions were present in higher concentration.

- Fort Thomas WTP – The DOC concentration in the raw water at the Fort Thomas WTP averaged 1.8 mg/L during the testing period. The removals of DOC and UV254 absorbing materials by a MIEX® - coagulation sequence were adequate to reduce the THM and HAA5 formation after three days of chlorine contact time to less than their respective regulatory limits of 80 μg/L and 60 μg/L. However, the levels of THM formation were still close enough to the regulatory limit to require further study to confirm the potential for meeting goals at this location with this treatment sequence.

  MIEX® treatment alone was more effective in this plant in reducing the TTHM and HAA5 formation than coagulation alone. A significant benefit observed for the coagulation process was that effective coagulation could be achieved at ferric sulfate doses of 10 mg/L or less with MIEX® pretreatment as compared with a combined coagulation approach that presently applies 5 mg/L of ferric sulfate in combination with 15 to 30 mg/L of a commercial acidified alum- polymer blend. This suggests that the approach to coagulation could be simplified and the costs associated with coagulation could be reduced. The reduction in coagulant use would yield reductions in the quantity of residuals produced and the associated disposal cost.

- Alfred Merritt Smith WTP – The DOC concentration in the raw water at the AMSWTF was 2.34 mg/L. Removal of DOC, while significant, was less than 50 percent at this location. Even at this rate of removal, the MIEX® process was capable of significant reductions in THM formation to levels approaching the limit of 80 μg/L, but did not reduce the formation to levels that would meet this limit in the testing that was performed. Such reductions may still be valuable at this location due to the fact that this is a direct filtration facility with constraints for increasing the coagulant dose and adjusting pH to levels that would have a significant impact on DOC removal. Additional treatment steps may also act to provide an integrated approach to control of DBP formation. This could include operation in sequence with an ozone – biofiltration sequence as is presently being planned at this plant. These considerations further emphasize the need to examine the MIEX® process as a component of an integrated treatment sequence. A capability of the MIEX® process to improve the stability of ozone residuals was another potential benefit that was observed at this location

There was little bromide removed by the MIEX® process, even when a low bromide salt was used in preparing the regenerant brine. This can probably be attributed to high concentrations of competing anions in the source water as indicated by the high concentrations of TDS and sulfate.

A potential for adverse effect on the rate of head loss buildup in the direct filtration process was observed. This effect is probably a consequence of the lack of a clarification step to remove particulates that are added by the MIEX® process.

Study results indicate that the MIEX® process can have a range of potential applications to meet site-specific needs. The costs of the process would generally be greater than those associated with treatment changes such as optimizing coagulation for greater removal of NOM in a conventional treatment sequence or conversion to chloramines as an alternative disinfection approach to limit the formation of THMs and HAAs. Many utilities may elect to consider these types of changes as first priorities before considering the MIEX® process. However, these approaches may not always be adequate. In addition, coagulation processes may not be available in all locations or may have constraints that limit capability for further optimization such as would be the case for many direct filtration facilities. There is also a degree of reluctance to rely on chloramines at some locations and this approach is not always available. Where applicable, the MIEX® process may prove to be less costly for NOM removal than approaches such as GAC or nanofiltration and could offer utilities a reasonable alternative to these technologies for situations where added levels of NOM removal are desired. Achievable levels of NOM removal may be greater with these other types of technologies and situations could occur where the MIEX® process may not provide the removals that are desired or where these technologies may prove to be more cost effective. However, the levels of NOM removal that are available with the MIEX® process may be appropriate at less cost to meet the needs of many utilities. Other alternatives that may become considerations in a larger disinfection and disinfection byproduct planning context are alternative oxidant/disinfection approaches, such as ozone, chlorine dioxide, and membrane filtration. In some cases, the MIEX® process might be used in combination with some of these processes. Therefore, a range of possibilities exist. Costs and process applicability of the respective process alternatives will vary depending on site specific factors. Costs that relate to disposal of regenerant in particular could exhibit significant variation for the MIEX® process from one site to the next. Benefits of the MIEX® process in relation to other plant processes will also be site specific. These benefits can include reductions in coagulant dose, solids to be disposed, and oxidant/disinfectant doses, and possible reductions in capital costs of downstream processes. These benefits will be considerations in computing the net cost impact of adding the MIEX® process.

The MIEX® process can achieve significant levels of removal of NOM and reductions in DBP formation, but the extent of these effects varies with water quality and it is important to confirm applicability with pilot testing. Bench scale testing can augment pilot testing, but caution should be exercised in using resin that has not been conditioned in a flow through sequence as the full-scale process relies on resin that has reached a steady state condition. Testing with virgin resin will not reflect removals that can be expected as a consequence. In many cases, the MIEX® process will be integrated within a treatment sequence and will provide a component of the overall strategy for NOM removal and DBP control. It is important that the process be examined for its capability to act in conjunction with other processes to meet the overall goals for NOM removal and DBP control under these conditions.

REFERENCES

Aiken, G.R. 1985. *Humic Substances in Soil, Sediment and Water*. Wiley-Interscience, New York.

Aiken, G.R., D.M. McKnight, K.A.Thorn, and E.M. Thurman, 1992. Isolation of Hydrophilic Organic Acids from Water Using Nonionic Macroporous Resins. *Org. Geochem*, 18(4):567.

Aiken, G.R., Leenheer .J.L. 1993. Isolation and Chemical Characterization of Dissolved and Colloidal Organic Matter. *Chemistry and Ecology*, 8:135.

APHA, AWWA, and WEF (American Public Health Association, American Water Works Association, and Water Environment Federation). 1998 Standard Methods for the examination of Water and Wastewater. 20th ed. Washington, D.C.:APHA.

Amy, G.L., B.C. Alleman, and C.B. Cluff. 1990. Removal of Dissolved Organic Matter by Nanofiltration. *Jour. AWWA*, 116(1)200.

Anderson, C.T., Maier, W.J. 1979. Trace Organics Removal by Anion Exchange Resins. *Jour. AWWA*, 71(5):278.

Babcock, D.B. and P.C. Singer. 1979. Chlorination and Coagulation of Humic and Fulvic Acids. *Jour. AWWA*, 71(3):149.

Bellar, T.A., J.J. Lichtenberg, and R.C. Kroner. 1974. The Occurrence of Organo-halides in Chlorinated Drinking Water. *Jour. AWWA*, 66(12):703.

Black, B.D., G.W. Harrington, and P.C. Singer. 1996. Reducing Cancer Risks by Improving Organic Carbon Removal. *Jour. AWWA*: 40.

Boening, P.H., D.D. Beckmann, and V.L. Snoeyink. 1980. Activated Carbon Versus Resin Adsorption of Humic Substances. *Jour. AWWA*: 54.

Bolto, B., D. Dixon, R. Eldridge, and S. King. 2002. Removal of THM precursors by coagulation of ion exchange. *Wat. Res*, 36:5066.

Bolto, B., D. Dixon, R. Eldridge, S. King, and K. Linge. 2002. Removal of Natural Organic Matter by Ion Exchange. *Wat. Res*, 36:5057.

Bourke, M., M. Slunjski, B. O'Leary, and P. Smith. 1999. Scale-Up of the MIEX® DOC Process for Full Scale Water Treatment Plants. *Proc. 18th Fed. Conv. Aust. Water and Wastewater Assoc., Adelaide*.

Bourke, M., S. Harrison, B. Long, and T. Lebeau. 2001. MIEX® Resin Pretreatment Followed by Microfiltration as an Alternative to Nanofiltration for DBP Precursor Removal. *Membrane Technology Conference Proceedings*.

Bourke, M. 2001. Use a Continuous Ion Exchange Process (MIEX®) to Remove TOC and Sulfides From Florida Water Supplies. *Florida Water Resources Conference*.

Bourke, M., T. Champlin, J. Perella, C. Topham, and J. Blattman. 2002. Use of a Magnetic Ion Exchange Resin to Improve DBP Precursor Removal and Reduce Coagulant Usage at Lee County's Olga WTP. *Florida Section AWWA Annual Convention*.

Botica, C. Personal Communication March, 2004.

Bove, F., M. Fulcomer, J. Klotz, J. Esmart, E. Dufficy, and E. Savrin. 1995. Public Drinking Water Contamination and Birth Outcomes. *American Journ. Epidemiology*, 141:850.

Brattebo, H., H. Odegaard, and O. Halle. 1987. Ion Exchange for the Removal of Humic Acids in Water Treatment. *Wat. Res*, 21(9):1045.

Brophy, K.S., H.S. Weinberg, and P.C. Singer. 2000. Quantification of Nine Haloacetic Acids Using Gas Chromatography with Electron Capture Detection. *ACS Symposium Series*, ACS Symposium Series 761:425.

Cantor, K.P. 1997. Drinking Water and Cancer. *Cancer Causes and Control*, 8:292.

Cantor, K.P., C.F. Lynch, M.E. Hildesheim, M. Dosemeci, J. Lubin, M. Alavanja, and G.F. Craun. 1998. Drinking Water Source and Chlorination By-Products in Iowa. *Epidemiology*, 9:21.

Chang, E.E., Y.P. Lin, P.C. Chiang. 2001. Effects of Bromide on the Formation of THMs and HAAs. *Chemosphere*, 43:8:1029.

Chin, Y.; G. Aiken and E. O'Loughlin. 1994. Molecular Weight , Polydispersity, and Spectoscopic Properties of Aquatic Hhumic Substances. *Env. Sci. & Tech*, 28(11):1853.

Christman, R.F., D.S. Norwood, D.S. Millington, J.D. Johnson, and A.A. Stevens. 1983. Identity and Yields of Major Halogenated Products of Aquatic Fulvic Acid Chlorination. *Env. Sci. & Tech*, 17(10):625.

Collins, M.R., G.L. Amy, and P.H. King. 1985. Removal of Organic Matter in Water Treatment. *Journ. Environ. Eng.*, 111(6):850.

Collins, M.R., G.L. Amy, G.L, and C. Steelink. 1986. Molecular Weight Distribution, Carboxylic Acidity, and Humic Substances Content in Aquatic Organic Matter: Implications for Removal During Water Treatment. *Env. Sci. & Tech.*, 20(10):1028.

Cook, D., C. Chow, and M. Drikas. 2001. Laboratory Study of Conventional Alum Treatment Versus MIEX® Treatment for Removal of Natural Organic Matter. *19th Federal AWWA Convention Proceedings.*:AWWA.

Cooper, W.J., G.L Amy, C.A. Moore, and R.G. Zika. 1986. Bromoform formation in ozonated groundwater containing bromide and humic substances. *Ozone Sci Eng*, 8(1):63.

Couri, D.M., S. Abdel-Rahman, and R.J. Bull. 1982. Toxicological effects of chlorine dioxide chlorite, and chlorate. *Environ. Health Perspectives*:46:13.

Cowman, G. A., and P.C. Singer. 1996. Effect of Bromide Ion on Haloacetic Acid Speciation Resulting from Chlorination and Chloramination of Aquatic Humic Substances. *Env. Sci. & Tech*, 30(1):16.

Croue, J.P., J.F. Debroux, G. Aiken, J.A. Leenheer and G.J. Amy. 1999. Natural Organic Matter: Structural Characteristcs and Reactive Properties. *Formation and Control of Disinfection By-Products in Drinking Water.*

Danielesson, L.G. 1982. On the Use of Filters for Distinguishing Between Dissolved and Particulate Fractions in Natural Waters. *Wat. Res*, 16:179.

Delphos, P., K. Harr and T. Bianco. 2001. Selective DBP Precursor Removal with an Innovative Ion Exchange Process. *Water Quality Technology Conference Proceedings.*:AWWA

Dodds, L., W. King, C. Woolcott, and J. Pole. 1999. Trihalomethanes in Public Water Supplies and Adverse Birth Outcomes. *Epidemiology*, 10(3):233.

Drikas, M. 2003. Operating the MIEX® Process with Microfiltration or Coagulation. *Proceedings of the 2003 Water Quality and Technology Conference, November.* Philadelphia, PA.:AWWA.

Edzwald, J.K., J. McIntyre, R.L. Sanks, M.J. Semmens and J.S. Taylor. 1979. Organics Removal by Coagulation: A Review and Reseach Needs. *Jour. AWWA*, 71(10):588.

Edzwald, J.K., W.C. Becker, and K.L. Wattier. 1985. Surrogate Parameters for Monitoring Organic Matter and THM Precursors. *Jour. AWWA*, 7(4):122.

Edzwald, J.K. 1993. Coagulation in drinking water treatment: particles, organics and coagulants. *Water Sci and Technol*, 27(11):21.

Farahbakhsh, K., S.S. Adham, and D.W. Smith. 2003. Monitoring the Integrity of Low-Pressure Membranes. *Jour. AWWA*, 95(6):95.

Fettig, J. 1999. Removal of Humic Substances by Adsorption/Ion Exchange. *Wat. Sci. Tech.*, 40(9):173.

Fu, P.L.K., and J.M. Symons. 1990. Removal of Aquatic Organic Substances by Anion Exchange Resins. *Jour. AWWA:* 70.

Gallard, H. and U. von Gunten. 2002. Chlorination on Natural Organic Matter: Kinetics of Chlorination of of THM Formation. *Wat. Res*, 36(1):65.

Haag, W.R. and J. Hoigne. 1983. Ozonation of bromide containing waters: kinetics of formation of hypobromous acid and bromate. *Env. Sci. & Tech*, 17(5):261.

Jackson, P.E. 2000. *Ion Chromatography in Environmental Analysis*. John Wiley & Sons Ltd., Chichester.

Johnson, C.J. 2003. Impact of a Magnetic Ion Exchange Resin on Ozone Demand and Bromate Formation. *Department of Environmental Sciences and Engineering.*

Karanfil, T et al. 2003. Selecting Filter Membranes for Measuring DOC and UV_{254}, *Jour AWWA* 3:86.

Kargalioglu, Y., B.J. McMillan, R.A. Minear, and M.J. Plewa. 2000. *Natural Organic Matter and Disinfection By-Products: Characterization and Control in Drinking Water.* American Chemical Society, Washington, D.C.

Kavanaugh, M.C. 1978. Modified Coagulation for Improved Removal of Trihalomethane Precursors. *Jour. AWWA*, 11:613.

Kim, P.H.S. and J.M. Symons. 1991. Using Anion Exchange Resins to Remove THM Precursors. *Jour. AWWA*, 83(12):61.

King, W.D. and L.D. Marrett. 1996. Case-Control Study of Water Source and Bladder Cancer. *Cancer Causes and Control*, 7:596.

Ko Y.W., P.C. Chiang, and E.E. Chang. 1996. The Effect of Bromide Ion on the Formation of Organohalogen Disinfection By-Products During Ozonation. *Ozone Sci Eng*, 18(4):349.

Krasner, S.K., M.J. McGuire, and J.J. Jacangelo, 1989. The Occurrence of Disinfection By-Products in US Drinking Water. *Jour. AWWA*, 81(8):41.

Krasner, S.K., M.J. Sclimenti, and E.G. Means. 1994. Quality Degradation: Implications for DBP Formation. *Jour. AWWA*, 86(6):34.

Krasner, S.W., and G. Amy. 1995. Jar-Test Evaluations of Enhanced Coagulation. *Jour. AWWA*, 10:93.

Krasner, S.K., J.P. Croue, P.A. Hacker, C.J. Hwang, C.Y. Kuo, A.A. Modfidi, and M.J. Sclimenti. 1996. Three Approaches for Characterizing NOM. *Jour. AWWA*, 88(6):66.

Kroll, R.B., G.D. Robinson, J.H. Chung, 1994. Characterization of Trihalomethane (THM)-Induced Renal Dysfunction in the Rat .2. Relative Potency of THMs in Promoting Renal Dysfunction. *Archives of Environmental Contamination and Toxicology,* 27(1):5-7.

Lange. R., J. Tattersall, C. Botica, and P. Smith. 2001. Design and Construction of World's First Large Scale MIEX® Water Treatment Plant. *Proceedings of 19th Federal Convention, Australian Water & Wastewater Association*, Canberra, Australia. April.

Lee, N., S. Shahnawaz, G. Amy, and M. Bourke. 2003. Removal of Polar Natural Organic Matter (NOM) with a Magnetically Impregnated Ion Exchange (MIEX®) Media. *Proceedings of the 2003 American Water Works Association Annual Conference.* Anaheim, CA.: AWWA.

Liang, L. and P.C. Singer 2003. Factors Influencing the Formation and Relative Distribution of Haloacetic Acids and Trihalomethanes under Conctrolled Chlorination Conditions. *Env. Sci. & Tech*, 37(13):2920.

Linder, R.E.; G.R. Klinefelter, L.F. Strader, J.D. Suarex, and N.L. Roberts. 1997. Spermatotoxicity of Dichloroacetic Acid. *Reproductive Toxicology*, 11(5):681-688.

Long, B.W. Personal Communication. January 2004.

Malcolm, R.L., and P. McCarthy. 1992. Quantitative Evaluation of XAD-8 and XAD-4 Resins Used in Tandem for Removing Organic Solutes from Water. *Environ. International*, 18:597.

Masters, D. 2003 Wanneroo MIEX® Plant Process Performance Monitoring and Optimisation Study. Water Corporation of Western Australia Report. Sept. 2003.

Miles, A. 2002. Comparison of Trihalomethanes in Tap Water and Blood. *Env. Sci. & Tech*, 36(8):1692.

Morran, J., D. Bursil, M. Drikas, and H. Nguyen, 1996. A New Technique for the Removal of Natural Organic Matter. *Proceedings of 26th AWWA Water Quality Technology Convention*. Denver, CO.: AWWA.

Morris, R.D.; A.M. Audet, I.F. Angelillo. 1992. Chlorination, chlorination by-products and cancer: a meta-analysis. *American Journ Public Health*, 82(7):955.

Najm, I. and S. Krasner.1995. Effects of Bromide and NOM on By-Product Formation. *Jour. AWWA*, 87(1):106.

NCI. 1976. Report on the carcinogenesis bioassay of chloroform. NTIS PB-264-018, Bethesda, MD.

Obolensky, A. and P.C. Singer. 2003. Analysis of Halogen Substituted Patterns in DBPs Using ICR Data and Their Applications. *Proceedings of Water Quality Technology Conference*, Denver, CO.: AWWA.

O'Brien J. Personal Communication. January 2004.

Oliver, B.G. and E. M. Thurman. 1983. *Water Chlorination: Environmental Impact and Health Effects*. Ann Arbor Publishers.

Owen, D.M.; G.L. Amy, Z.K. Chowdhury, R. Paode, G. McCoy, and K. Viscosil. 1995. NOM Characterization and Treatability. *Jour. AWWA*:46.

Pegram, R.A.; M.E. Andersen, S.H. Warren, T.M. Ross, and L.D. Claxton. 1997. Glutathione S-Transferase-Mediated Mutagenicity of Trihalomethanes in Salmonella Typhimurium: Contrasting Results with Bromodichloromethane and Chloroform. *Toxicology and Applied Pharmacology*, 144 (1): 183-188.

Pelekani, C.; M. Drikas and D. Bursill. 2001. Application of Direct Filtration to MIEX® Treated River Murray Water. *Florida Water Resources Conference Presentation*.

Pourmoghaddas, H., A.A. Stevens, R.N. Kinman, R.C. Dressman, L.A. Moore, and J.C. Ireland. 1993. Effect of Bromide Ion on Formation of HAAs During Chlorination. *Jour. AWWA*, 1:82.

Randtke, S.J. 1988. Organic Contaminant Removal by Coagulation and Related Process Combinations. *Jour. AWWA*, 80(5):40.

Reckhow, D.A. and P.C. Singer. 1984. Removal of Organic Halide Precursors by Preozonation and Alum Coagulation. *Jour. AWWA*, 76(4):151.

Reckhow, D.A. and P.C. Singer. 1985. Mechanisms of Organic Halide Formation and Implications with Respect to Pre-ozonation. *Water Chlorination: Environmental Impact and Health Effects*.

Reckhow, D.A., P.C. Singer, and R.L. Malcolm. 1990. Chlorination of humic materials: by-product formation and chemical interpretations. *Env. Sci. & Tech*, 24(11):1655.

Reif, J.S., M.C. Hatch, M. Bracken, L.B. Holmes, B.A. Schwetz, and P.C. Singer. 1996. Reproductive and Developmental Effects of Dinsinfection By-Products in Drinking Water. *Environ. Health Perspectives*, 104:1056.

Rook, J.J. 1974. Formation of Haloforms During Chlorination of Natural Waters. *Water Treatment Exam*, 23(2):234.

Rook, J.J. 1976. Haloforms in Drinking Water. *Jour. AWWA*, 68(3):168.

Rook, J.J., and S. Evans. 1979. Removal of Trihalomethane Precursors from Surface Waters Using Weak Base Resins. *Jour. AWWA*:71:520.

Rowland, M.D., Personal Communication (2004).

Schmitt D., F. Saravia, F.H. Frimmel, and W. Schuessler. 2003. NOM-facilitated Transport of Metal Ions in Aquifers: Importance of Complex-Dissociation Kinetics and Colloid Formation. *Water Research*, 37(15):3541.

Semmens, M., M. Burckhardt, D. Schuler, P, Davich, M. Slunjski, M. Bourke, and H. Nguyen. 2000. An Evaluation of Magnetic Ion Exchange (MIEX®) for NOM Removal. *Proceedings of 2000 Annual AWWA Conference*. Denver, Colo.: AWWA

Singer, P.C., J.J. Barry, G.M. Palen, and A.E. Scrivner. 1981. Trihalomethane formation in North Carolina drinking waters. *Jour. AWWA*, 73(8):392.

Singer, P.C. 1993. Disinfection By-Products in U.S. Drinking Waters: Past, Present, and Future. *National Conference on Environmental Engineering*:727.

Singer, P.C. 1994. Control of Disinfection By-Products in Drinking Water. *Journal of Environmental Engineering*, 120(4):727.

Singer, P.C. 1999. Humic Substances as Precursors for Potentially Harmful Disinfections By-Products. *Wat. Sci. Tech.*, 40(9):25.

Singer, P.C. and K. Bilyk. 2002. Enhanced Coagulation Using a Magnetic Ion Exchange Resin. *Water Research*.

Slunjski, M., M. Bourke, H. Nguyen, M. Ballard, J. Morran, and D. Bursill. 1999. MIEX® DOC Process - A New Ion Exchange Process. *Proceedings of the 18th Federal Convention*.

Slunjski, M., M. Bourke and B. O'Leary. 2000. MIEX® DOC Process for Removal of Humics in Water Treatment. *IHSS - Australian Chapter Seminar*.

Slunjski, M., H. Nguyen, M. Ballard, R. Eldridge, J. Morran, M. Drikas, B. O'Leary, and P. Smith. 2002. MIEX® - Good Research Commercialized. *Water*, 29(2):42.

Smith Paul, C. Botica, B. Long and B. Allender. 2002. Design, Construction, Commissioning and Operation of the World's First Large Scale MIEX® Water Treatment Plant. *Proceedings of 2002 Annual Conference, AWWA,* New Orleans, LA. June 2002.

Smith, Paul. Personal Communication. February 2003.

Stevens, A.A.; C.J. Slocum, D.R. Seeger, and G.G. Robeck. 1976. Chlorination of Organics in Drinking Water. *Jour. AWWA*, 68(11):615.

Summers, R.S., S.M. Hooper, H.M. Shukairy, G. Solarik, and D. Owen, 1996. Assessing DBP Yield: Uniform Formation Conditions. *Jour. AWWA*, 88(6):80.

Symons, J.M.; P. Fu.; P. Kim. 1995. Sorption and Desorption Behavior of Natural Organic Matter onto Strong-Base Anion Exchangers. In: Ion Exchange Technology. Ed: A. Sengupta, Lancaster, PA, Technomic Publishing Co.

Symons, J.M., S.W. Krasner, L.A. Simms, and M. Sclimenti. 1993. Measurement of THM and Precusor Concentrations Revisited: The Effect of Bromide Ion. *Jour. AWWA*, 93(1):1.

Thurman, E.M., and R.L. Malcolm. 1981. Preparative Isolation of Aquatic Humic Substances. Environmental Science and Technology, 15(4):463.

USEPA (United States Environmental Protection Agency). 1979. National Interim Primary Drinking Water Regulations; Control of Trihalomethanes in Drinking Water. Federal Register 63 (94) 69390-69476.

USEPA (United States Environmental Protection Agency). 1995. Methods for the Determination of Organic Compounds in Drinking Water-Supplement III. EPA/600/R-95-131 and EPA/600/R-95-131. Cincinnati, OH.

USEPA (United States Environmental Protection Agency). 1997. Method 300.1, Cincinnati, OH.

USEPA. (United States Environmental Protection Agency). 1998a. National Primary Drinking Water Regulations: Interim Enhanced Surface Water Treatment; Final Rule. Federal Register, 63(241).

USEPA (United States Environmental Protection Agency). 1998b. National Primary Drinking Water Regulations: Disinfectants and Disinfection By-Products, Final Rule. FR 98-32887, Washington, DC.

USEPA (United States Environmental Protection Agency). 2003. National Primary Drinking Water Regulations: Stage 2 Disinfectants and Disinfection By-products Rule; Proposed Rule. Federal Register, 68(159).

vonGuten, U., and J. Hoigne. 1992. Factors Controlling the Formation of Bromate During Ozonation of Bromide-Containing Waters. *Aqua*, 42(5):299.

Waller, K.S.H.S., G. DeLorenze, and B. Hopkins. 1998. Trihalomethanes in Drinking Water and Spontaneous Abortion. *Epidemiology*, 9(2):134.

Werdehoff, K.S., and P.C. Singer. 1987. Chlorine dioxide effects on THMFP, TOXFP, and the formation of inorganic by-products. *Jour. AWWA*, 79(9):107.

Wershaw, R.L. and G.R. Aiken. 1985. Molecular weight and size measurements of humic substances. In Humic Substances in Soil, Sediment, and Water – Geochemistry, Isolation, and Characterization. New York; John Wiley and Sons, p. 477-492.

White, M.C., J.D. Thompson, G.W. Harrington, and P.C. Singer. 1997. Evaluating Criteria for Enhanced Coagulation Compliance. *Jour. AWWA*, 89(5):64.

Wolfe, R.L. 1990. Ultraviolet disinfection of potable water. *Env. Sci. & Tech*, 24(6):768.

Zegers, R.E., D.J. Rexing, E.C. Wert, T. Glew, J.J. Neemann, R.A. Hulsey, B.W. Long. 2001. Bromate Formation and Mitigation While Achieving 2-log *Cryptosporidium* Inactivation in Lake Mead Water. *Proceedings of the 2001 American Water Works Association Annual Conference*. Washington, D.C. June 12-16, 2001. Denver, Colo.: AWWA.

Zhang, X. and R. Minear. 2002. Decomposition of trihaloacetic acids and formation of the corresponding trihalomethanes in drinking water. *Wat. Res*, 36:3665.

ABBREVIATIONS

ACS	American Chemical Society
AMSWTF	Alfred Merritt Smith Water Treatment Facility
APHA	American Public Health Association
AWWA	American Water Works Association
AwwaRF	Awwa Research Foundation
BIF	bromine incorporation factor
°C	degrees Celsius
$CaCO_3$	calcium carbonate
CFU	colony forming unit
cm	centimeters
CSIRO	Commonwealth Scientific and Industrial Research Organisation
DBP	disinfection by-product
DOC	dissolved organic carbon
ECD	electron capture detector
EPA	USEPA method
°F	degrees Fahrenheit
ft	foot
ft^2	square foot
FTWTP	Fort Thomas Water Treatment Plant
g	gram
GAC	granular activated carbon
gpm	gallons per minute
gpm/sf	gallons per minute per square foot
HAA	haloacetic acids
HAA5	five haloacetic acid compounds analyzed
HAA9	nine haloacetic acid compounds analyzed
HCl	hydrochloric acid
HPC	heterotrophic plate count
HPIA	hydrophilic acids
HPOA	hydrophobic acid
HWTP	Hanahan Water Treatment Plant
IS	internal standard
L	liter
lb	pound
LGW	laboratory grade water

L/min	liters per minute
L/ML	liter per million liters
MCL	maximum contaminant level
mg/L	milligrams per liter
mgd	million gallons per day
m/h	meter per hour
min	minute
mL	milliliter
mM	millimole
ML/d	million liters per day
MtBE	methyl tertiary butyl ether
MWTP	Maytum Water Treatment Plant
μg/L	micrograms per liter
μm	micrometers
NA	not applicable
NaCl	sodium chloride
NaOH	sodium hydroxide
NF	nanofiltration
nm	nanometer
NMW	nominal molecular weight
NOM	natural organic matter
NSRS	non-sulfide-reduced-sulfur
ntu	nephelometric turbidity unit
PAC	powdered activated carbon
psi	pounds per square inch
PTFE	polytetrafluoroethylene
rpm	rotations per minute
sec	seconds
SM	Standard Method
SUVA	specific ultraviolet absorbance
TDS	total dissolved solids
TFE	tetrafluoroethylene
THM	trihalomethane
TTHM	total trihalomethane
TOC	total organic carbon
TOX	total organic halide
TPHA	transphilic acids
UFC	uniform formation conditions

UNC	University of North Carolina Chapel Hill
USEPA	United States Environmental Protection Agency
UV254	ultraviolet light having a wave length of 254 nanometers
WGTP	Wanneroo Groundwater Treatment Plant
WTP	Water Treatment Plant
X_2AA	di-halogenated acetic acids
X_3AA	tri-halogenated acetic acids

IWA Publishing's authorised EU representative for General Product Safety Regulations is Diane D'Arras, 15 rue Duret, 75116 Paris, France, e-mail: safety@iwap.co.uk.

Printed and bound by CPI Group (UK) Ltd, Croydon, CR0 4YY
17/09/2026
02218811-0017